Clinical Core Laboratory Testing

Ross Molinaro • Christopher R. McCudden
Marjorie Bonhomme • Amy Saenger
Editors

Clinical Core Laboratory Testing

Editors
Ross Molinaro
Emory Medical Laboratories
Emory University School of Medicine
Atlanta, GA, USA

Marjorie Bonhomme
David Geffen School of Medicine
University of California at Los Angeles
Los Angeles, CA, USA

Christopher R. McCudden
University of Ottawa
Ottawa, ON, Canada

Amy Saenger
Mayo Clinic
Rochester, MN, USA

ISBN 978-1-4939-7958-5 ISBN 978-1-4899-7794-6 (eBook)
DOI 10.1007/978-1-4899-7794-6

Printed on acid-free paper

This Springer imprint is published by Springer Nature
The registered company is Springer Science+Business Media LLC
The registered company address is: 233 Spring Street, New York, NY 10013, U.S.A.

To James J. Miller (Jim), whose contributions to the field of clinical laboratory medicine will live on through his writings and the people he influenced.

Foreword

My father is 63 years old and has always been in great health. He is an enthusiast of sports and continues to play tennis with several friends at least twice a week. When we were young, he introduced us to tennis and badminton. As a weekend pastime, our home's backyard was large enough to fit a badminton court and, together with my brother and sister, we made teams to play either doubles or round robin. In the past years, the badminton "tournaments" have become much more complex as grandchildren want to participate. Some of these "tournaments" can become quite heated discussions regarding the formation of the teams and who the "winner" is. As you can imagine some of the most interesting "tournaments" occur while everyone is visiting my parents during holidays. Last year at Thanksgiving, we were in the heat of the "tournament" when my father started complaining about severe pain in the middle of his left leg. It did not appear to be a sprain since the pain was in the middle of the thigh. The pain was intense enough and he seemed out of breath so he had to stop playing.

The next morning he continued to have leg pain and the area was swollen and red. Even though my father did not want to go to the hospital, my sister and I insisted on taking him. When we arrived at the emergency room of the hospital, there were a lot of patients. After waiting for about one hour, we were passed into a room where a young doctor took a short history, looked at my father's leg, and examined his heart and lungs. She told us she needed to get an X-ray and some blood work before deciding how to proceed. I asked her what the blood was for and she told me that she wanted to check if everything was normal and to count different cells that were in my father's blood. A few minutes later another person came into the room, introduced herself as the phlebotomist, and proceeded to explain that she was going to draw some blood for the tests the doctor had ordered. I asked her if she was the one that was going to test the blood or if this was going to be done by the young doctor that had seen my father a few minutes before. She told me that neither she nor the doctor would be doing the testing, as this was done by instruments in the laboratory, which are operated by medical technologists. I asked the phlebotomist first what a medical technologist was, and second, how long would it take for the results to come back. As we were in this conversation, the orderlies that were taking my father to the X-ray came into the room. The phlebotomist drew the blood and my father was taken away.

The conversation about looking and counting blood cells left me interested. While waiting for my dad to come back from X-ray, I went to the waiting area and used my phone to investigate further. Using the web I learned that laboratories are in pretty much every hospital but not much is described as to how they work. I've watched CSI and thought that may be what it is like. Being an engineer myself, I was fascinated by how physics, chemistry, and engineering have been the basis of how the lab may use these aspects of science to let the doctor know what was wrong with my dad.

As my father came back from the X-ray, the doctor came in and explained that he had fractured his femur, that the fracture had occurred because he had some lesions in the bone that made the bone soft, and that there were other of these "lytic" lesions in other areas of the bone. She said that for the time being she needed to immobilize the leg and that my father needed to see two different specialists as soon as possible. One of the specialists would work on the fracture while the hematologist would further study why he had these lesions. She said that people in the emergency room would help us do the appointments and that in the mean time she would take some more blood for further studies so that when we had the appointment with the other doctors we would already have some tests that would help define the diagnosis. I asked again for the name of the lesions in the bone and she spelled the name for me "l-y-t-i-c." My sister asked if she already had the results from the blood tests and why could they not use the blood they had already drawn. The doctor said that the study had shown that my dad had anemia and that the red blood cells were stuck to one another because there was too much protein. She also commented that for the new tests a different kind of tube was necessary. I asked what other tests she was planning on doing and the doctor said that it was important to know which protein was increased and made the red blood cells stick to each other. My father asked the doctor what all this meant, if there was a particular diagnosis that she was thinking about. She said that at this point any particular diagnosis was premature, as there were a variety of diseases that are associated with lytic lesions in the bone including rare benign diseases to cancer.

After placing a brace on my father's leg, drawing more blood and making the appointments with the specialists for the next week, we left the emergency room. On our way home, we talked, trying to decide what we should tell my mother and the rest of the family. I had looked up on the web for causes of lytic lesions in the bones and found that at my father's age metastasis from a cancer or a blood malignancy called multiple myeloma was at the top, but a laboratory test is needed to make the diagnosis.

The day of our doctor's appointment, the doctor was very nice and explained what the laboratory tests had shown and others that would be needed. Different things, such as calcium measurements and more tests looking at the protein in my father's blood, were needed. What was strange to me was that the doctor said that more tests needed to be sent to the laboratory to further define exactly the disease process. He said that it may be multiple myeloma (a cancer of the cells that produce a specific type of protein called immunoglobulins).

I had never realized how much doctors need the laboratory to define different diseases. Up to then, the doctors that had seen my dad had needed two samples of blood in the emergency room, another sample of blood to accompany the sample from the bone marrow that was being obtained, and he had to collect urine for an entire day. All these samples were going to a laboratory where a variety of instruments and people that, to my knowledge, had never seen my father would be looking at the different proteins and cells in his blood. The people in the laboratory were an integral part of the diagnosis and treatment my father would receive, and I had never met them or heard about them before. This left me wondering where the lab was and how it worked.

Professor
Pathology & Laboratory Medicine
Emory

Jeannette Guarner

Preface

The clinical laboratory provides a vital service and significantly contributes to patient care. It is commonly cited that ~70 % of medical decisions are based on data and interpretive information generated from the laboratory. While this number may be slightly inflated, it is well accepted that a predominant number of critical decisions made by physicians on a daily basis are centered on laboratory test results.

Clinical laboratories have been transformed over the past 20 years from being slow and manual to highly automated, high-throughput production-type environments. At the same time, the amount of hands-on experience and time physicians spend in the lab has decreased dramatically. While physicians once played an active role in actually testing patient samples, it is now common for clinicians to never set foot in the laboratory. This disconnect has occurred at a time where the volume and complexity of information generated by the laboratory has grown exponentially.

Comprehensive medical student and resident education regarding operation and function of the clinical laboratory is limited and often overlooked due to time limitations of the trainees, who are also attempting to learn many areas of medicine. Furthermore, the amount of material bombarding care providers at all stages in their careers (medical student, resident, fellow, or practicing physician) is immense, and leaves little time to devote to learning the intricacies of laboratory medicine, despite the fact there is heavy reliance on laboratory results to effectively treat patients. Due to the growing complexity of diagnostic testing, it is important that care providers develop a better understanding of how the laboratory works. This will ultimately lead to improved laboratory service and patient care, better communication between medical disciplines, value-added improvements in test utilization, and ultimately a reduction in healthcare costs. The patient is also an integral part of care decisions and often assumes an active role in educating themselves and acting as an effective advocate. Patients may benefit from having a more comprehensive understanding of how the laboratory works to answer questions such as "What does a given lab result tell you?" "What does a 'laboratory error' really mean?" "Is the lab anything like how it appears on television?" The intent of this book is to dispel some common myths about the clinical laboratory and provide tangible, practical information that can help physicians and other medical providers utilize the laboratory more effectively. It is also intended to foster appreciation for the laboratory amongst healthcare providers, advocates, and patients so they can seek expertise from the laboratory when questions arise.

Early chapters focus on the structure and function of the clinical laboratory. The organization and scope of services that a laboratory provides depend heavily on the size of the lab. Labs can range in size from small/limited service physician office testing which may have one person drawing blood, collecting urine samples, and performing the testing to large national reference laboratories with dedicated aircraft available to deliver samples throughout the country or around the world. In between these two extremes lies the prototypical tertiary care center hospital laboratory at an academic medical center where both physicians and laboratory staff are trained and educated. Regardless of the size of the laboratory, there are frequently subspecialties and areas of expertise and different operational structures/leadership may exist within the laboratory. Knowledge of who has expertise in a various area can help focus problem solving and make communication more effective. As one may not seek care from a neurologist for a broken leg, one should not necessarily seek out a microbiologist for interpretation of cardiac marker results. Hospital laboratories often have separate testing areas for immunology, microbiology, cytology, hematology, biochemistry, and genetics (collectively termed "clinical pathology").

Clinical pathology operates completely differently than the area of pathology which interprets tissue specimens, termed anatomic pathology. Thus an appreciation for the scope of laboratory testing helps facilitate effective communication, realistic expectations for how long it takes to perform a certain test and if a given test can even be ordered. Should one expect a family practice clinic to provide rare genetic testing on site? Regulatory requirements follow the complexity of the laboratory and must be met in order to legally operate the laboratory and bill for services. The early chapters in this book also provide a concise outline of the scope, structure, and regulatory requirements that laboratorians face. In addition, discussion will focus around the general organizational hierarchy of the laboratory, which helps determine who should be consulted depending on the nature of the question or problem.

It is a common perception that hospital laboratories look and operate like a scene out of the TV show "CSI." The illusion of translucent computer screens and complete genome results in less than 5 min make for a satisfying story to a large audience, but, while these science fiction depictions of laboratory operations may become truth in the years to come, they have successfully intrigued the public and increased interest and awareness about different types of laboratories. The clinical laboratory, however, has not yet achieved that level of transparency, which can lead to false information and expectations from physicians and patients. The later chapters of this handbook focus on the role laboratories play in patient care by advocating for proper utilization through test ordering and result interpretation. These chapters will discuss what can go wrong in the lab testing process, and more importantly, some details of the where, when, and why of how it happens. Collectively, the objective of this book is to help those who rely on laboratory results to maximize the utility of laboratory services.

Ottawa, ON, Canada Christopher R. McCudden
Atlanta, GA, USA Ross Molinaro

Contents

Chapter 1
Laboratory Structure and Function

Roger L. Bertholf

Although its origin is obscure and many variations exist, writers and speakers commenting on laboratory quality often recite the phrase: *The right result, on the right test, on the right patient, at the right time*. It is a catchy phrase that captures the four most important considerations in medical laboratory services: integrity of laboratory results, efficient use of laboratory tests by medical staff, safeguards against patient identification errors, and timeliness. One might add *at the right price* to this list, since the economics of healthcare delivery are important considerations in clinical laboratory services, as well (discussed in Chap. 2). The importance of laboratory tests in the diagnosis and medical management of patients is widely acknowledged. Some have attempted to quantify the laboratory contribution towards medical care, often citing a statement attributed to Rodney Forsman, the Administrative Director and Assistant Professor of Laboratory Medicine and Pathology at Mayo Clinic College of Medicine in Rochester, MN, that 70 % of all medical decisions are based on laboratory data. Forsman's assertion may be true, but whether clinicians make 70 % of their decisions based on laboratory data, regard 70 % of laboratory results as contributing to their decisions, or assign a 70 % weight laboratory tests in their decision-making process is debatable, and mostly superfluous. Results of laboratory tests are elements in the equation that determines diagnosis and treatment. In some cases, they may dominate the equation. In other cases, their contribution may be minimal or nonexistent. Or perhaps more commonly, laboratory tests provide information that makes subtle changes in the trajectory of clinical decisions, sometimes lending support to the initial clinical impression, and other times revealing medical conditions that are not clinically evident.

R.L. Bertholf (✉)
Department of Pathology and Laboratory Medicine, University of Florida Health Science Center, 655 West 8th Street, Jacksonville, FL 32209, USA
e-mail: roger.bertholf@jax.ufl.edu

R. Molinaro et al. (eds.), *Clinical Core Laboratory Testing*,
DOI 10.1007/978-1-4899-7794-6_1

The right result. "Quality is Job One" was the advertising slogan adopted by Ford Motor Company in 1982, in response to the perception that Japanese and European auto manufacturers had surpassed American made brands in the quality of their products. The slogan has always been relevant to clinical laboratories. The integrity of test results is the foremost priority for every medical laboratory professional. To ensure quality, surveillance programs are put in place to monitor the performance of analytical methods. Quality control (also discussed in Chap. 4), always the staple of good laboratory practice, has evolved from a process that included controls with each batch of specimens to a periodic assessment at regular intervals, typically daily or every 8 h, since modern automated methods used in clinical laboratories have greater stability than the manual methods that preceded them. Newer approaches to quality control, currently being incorporated into regulatory and accreditation standards, may reduce the frequency even further. A key concept that emerged was standardization, which can be assessed through inter-laboratory agreement. Inter-laboratory agreement between analytical methods, which at one time was poor, has been greatly improved by the widespread implementation of proficiency testing (in Europe, Canada, and Australia these programs are called "external quality assessment," or EQA). The idea of inter-laboratory harmonization was hatched by the eminent clinical pathologist Dr. F. William Sunderman in the 1930s. His strategy for periodic proficiency challenges was codified into federal regulations when the Clinical Laboratory Improvement Act of 1967 was enacted. When the legislation was revised in 1988, with the Clinical Laboratory Improvement Amendments, proficiency testing was overhauled, and the requirements became stricter. Three rounds of proficiency challenges per year were required, and successful performance on 4 of 5 (80 %) challenges in any round for regulated analytes was required. In addition, successful performance on two of any three consecutive rounds was required. In some cases, 100 % pass rates are required, such as the case in blood banks where blood products are given to patients. Failure is not allowable in these cases since giving the wrong type blood to a patient may lead to catastrophic problems and even death. Laboratories that failed proficiency testing were subject to suspension from Medicare reimbursement, and failure to incorporate proficiency challenges into the normal workflow, involving all testing personnel and prohibiting any procedure that would subvert the intent for proficiency testing to reflect the quality of routine testing, were subject to criminal penalties. Clinical laboratories adapted to these requirements, and also developed quality assurance programs that addressed other aspects of quality such as reference interval validation and periodic assessment of linearity in quantitative methods.

The right test. Overutilization of laboratory services increases the cost of healthcare. In the USA, few checks exist on the patterns clinicians adopt for ordering laboratory tests on their patients. "Medical necessity" rules were implemented in the 1990s, requiring that certain clinical criteria be met when a particular laboratory test is ordered, and also requiring that patients consent to the laboratory tests, with the understanding that they may not be reimbursed. The traditional unfocused "shotgun" approach to laboratory assessment was curtailed, and disease-specific laboratory profiles were created. A laboratory test needed a reason to be ordered, and patients needed to be informed that they might be financially responsible for unjustified laboratory tests. In principle, it was a genuine effort to limit the health-care costs associated with laboratory testing, but in practice, there is little evidence it has reduced the cost of healthcare. This may be the biggest challenge to clinical laboratories, to promote more efficient use of laboratory resources. Laboratory directors and administrators focus on the most efficient ways to provide services, but often neglect the advantages to be gained by more prudent utilization (see Chap. 2 for more on test utilization).

On the right patient. Errors occur in all areas of medicine and the clinical laboratory is not immune to these faults. Some are analytical, some may be due to misinterpretation of a laboratory result, and some are due to improper collection of the specimen submitted for laboratory analysis. Many errors, however, are due to misidentification of the patient. Patient identification is one of the goals identified by The Joint Commission as essential to patient safety. Data on the frequency of patient identification errors suggest that it is a significant problem that may be underestimated because methods to detect mislabeled specimens, such as delta checks (discussed in Chap. 3), lack the sensitivity to reliably identify specimens that have been labeled with the wrong patient ID. Specimen labeling at the point of collection often is beyond the control of the laboratory unless phlebotomy services are within the domain of laboratory administration. Engineering controls are available that eliminate the need to relabel a specimen once it reaches the laboratory, greatly reducing the risk of mislabeling by laboratory staff.

At the right time. The timeliness of laboratory results, particularly in hospital-based clinical laboratories, has always been the focus of great attention by laboratory administrators and clinical staff. Results of certain tests may be required for urgent medical interventions; other laboratory tests may be required before a patient can be admitted or discharged, and timeliness of laboratory results often has a greater impact on operational efficiency than the quality of medical care. But regardless of the reason, laboratories are under constant pressure to improve the turnaround time for test results. One consequence of this desire for immediate results is the development and proliferation of point of care (POC) tests, which now comprise a significant fraction of all laboratory testing. It is an example of an entire industry that is based on a single perceived benefit—rapid turnaround time—because the unit cost of POC tests is higher, and the analytical performance is poorer, than corresponding tests performed in the clinical laboratory. For tests performed in the central laboratory, however, a design that promotes rapid transportation of specimens to the lab, efficient processing of the specimens once they reach the lab, and analyzer throughput that avoids delays in reporting results are essential considerations towards the goal of providing the shortest turnaround times for laboratory results.

In clinical laboratory practice, no amount of operational efficiency can compensate for inaccurate test results; the analytical integrity of lab tests remains the first priority. In this chapter, however, we will focus on how to most efficiently deliver the "right result," with the assumption that proper attention has been given to selecting the appropriate analytical methods and ensuring their proper function with adequate surveillance (QC, PT, validation studies, etc.).

Preanalytical factors such as specimen collection, labeling, and handling prior to delivery to the lab are estimated to cause over half of all errors in laboratory results (Plebani, 2006). The laboratory has limited control over preanalytical errors, but specimen collection and receipt procedures typically address these sources of error.

The design of clinical laboratory services involves many components. This review will focus on the economics, layout, logistics, and overall design of clinical laboratory services.

Types of Clinical Laboratories

There is a diverse array of strategies for providing clinical laboratory services to physicians:

- *Physician office laboratories* (POLs) typically serve physicians in single or group practices, are located within the practitioners' office facilities, and provide limited laboratory services exclusively for their private patients.
- *Hospital laboratories* mostly serve inpatients and emergency departments, but often receive specimens from outpatient clinics too.
- *Referral laboratories* provide services that may be local, regional, or national in scope, and accept specimens from other laboratories or physicians' offices.
- *Research laboratories* offer esoteric tests that may be useful in certain clinical scenarios.

For decades, physicians have performed laboratory tests—microscopic examination of urine sediment, KOH preparations for fungal infections, blood smears for erythrocyte morphology, hemocytometric differentials, fecal parasites, and wet mounts of vaginal discharge for trichomonads—in their offices. When dry reagent technology was developed by Helen Free[1] at Miles Laboratories for measuring glucose in blood and urine, another simple means for laboratory assessment was available to physicians. Benchtop chemistry and hematology analyzers became available in the late 1970s, and were aggressively marketed to POLs. Until CLIA '88 was enacted, physician office laboratories were unregulated, but the new requirements created by the amendments were more than most POLs could satisfy. Microscopy was particularly affected by CLIA '88, which created a specific category of regulated laboratory procedures, "Provider-Performed Microscopy" (PPM), to address competency requirements for healthcare providers performing these diagnostic laboratory procedures.

[1] After graduating with a BS degree in chemistry from the College of Wooster in 1944, Helen Murray began working in the research department at Miles Laboratories under the direction of Dr. Alfred Free, whom she married in 1947. Helen Free is credited with the development of the dry reagent technology that was incorporated into many products marketed to clinical laboratories, including the Dextrostix, Uristix, Ketostix, Labstix, and a still-currently marketed urine dipstick product, Multistix. Helen Free was elected president of the American Association for Clinical Chemistry in 1990, and president of the American Chemical Society in 1993.

Medical care of hospitalized patients requires uninterrupted availability of clinical laboratory services. Hospital laboratories are required to provide urgent results for critical patients, routine tests for monitoring therapy, and esoteric tests for difficult diagnoses. Hospital laboratories serve many needs, usually functioning as POLs for on-site outpatient clinics, referral laboratories for remote clients, a STAT laboratory for the emergency department, and as a referral center to send out laboratory tests that are available only in research laboratories. The many demands on hospital laboratory services create challenges in the allocation of resources to providing both urgent and routine tests.

Most hospital laboratories offer services to remote clients, usually termed "outreach" work. Outreach laboratory services usually include providers within the healthcare network of the hospital, but may involve independent providers, too. In an environment of capitated reimbursement for hospitalized patients, outreach services have become an essential source of revenue for hospital-based laboratories, since many patients visiting private physicians' offices have insurance that pays for laboratory tests on a fee for service basis. The challenge for hospital laboratories is to balance their responsibility to provide services to inpatients with the operational demands of outpatient laboratory services that often require nontechnical staff to transport, receive, and register specimens from patients who are not already registered as hospital inpatients with accompanying insurance information. In addition, outreach programs based in hospital laboratories compete with referral laboratories, and often economies of scale exist which put hospitals at a disadvantage in that arena.

Consolidation of laboratory services is an effective strategy for minimizing costs. Referral laboratories have the advantage of scale; the cost per test decreases as volume increases. Regional referral laboratories used to be common, but many have been absorbed by national reference labs that can take advantage of greater volume to minimize the cost of individual tests, and provide services such as conveniently located collection stations that are too expensive for smaller laboratories to establish. Referral laboratories are high-volume operations that leverage economies of scale to offer services at competitive prices. Laboratory economics are discussed in the next section but it is important to note, when comparing the different types of clinical laboratory services, that certain costs are fixed and therefore the higher the volume of tests performed, the smaller the fixed cost of laboratory operations is as a component of overall laboratory expenses. In a low-volume laboratory, overhead expenses—utilities, management, amortization of capital investments in equipment, service contracts, and other fixed costs described in the next section—are a significant fraction of the cost of producing a laboratory result. As volume increases, the contribution of fixed expenses to the cost per test decreases. Referral laboratories can operate at a lower margin per test, and therefore have the capability to offer laboratory services at a lower incremental cost than smaller laboratories can afford.

Referral laboratories are essential services because some tests are too expensive to offer unless the volume is high (e.g., frequently measured tumor markers). The unreimbursed overhead expenses involved in performing some of these laboratory tests include quality control, proficiency testing, expensive instrumentation, and sometimes personnel with advanced skills. These tests can only be economically performed when the volume is sufficient to offset the cost of maintaining the service, and referral laboratories, which serve a large patient population, have the capability of offering tests that wouldn't be economically feasible for smaller labs.

Some diagnostic tests are only available from the research laboratories where they were developed. Esoteric laboratory tests provide diagnostic information that is not routinely offered by clinical laboratories. Although some of these tests are FDA approved, many have not yet received approval. For example, when molecular methods were developed to probe genetic information that was diagnostically useful, the technology for these tests was at first limited to research laboratories. Molecular diagnostics are now widely deployed in clinical laboratories, but many of these methods are not FDA approved; some fall into a category that has been given the name "Lab Developed Tests," or LDTs. Accrediting agencies are trying to keep up with the challenge of validating these new methods.

Clinical Laboratory Economics

History

Although federal funding of healthcare was debated by congress throughout the first half of the twentieth century, the Medicare program as it currently exists wasn't established until President Johnson signed H.R. 6675 into law in 1965, creating a federally funded healthcare insurance program for anyone over the age of 65. The legislation was included with Title XIX amendments to the Social Security Act, created in 1935. The same legislation created Medicaid, a federally subsidized state program providing healthcare insurance for low-income citizens who could not afford the expense of medical care or insurance. In 1965 healthcare in the USA for everyone over the age of 65, and for those whose income was below a certain threshold, was for the first time in history insured by either the federal or state government.

Medicare and Medicaid reimbursement of medical expenses coincided with a revolution in clinical laboratory technology (Chap. 5). Radioimmunoassay had been described only a few years earlier by Yalow and Bersen in 1960, and the ability to measure hormones transformed endocrinology into routine medical practice. The introduction of solid state microprocessors accelerated the development of automated chemistry analyzers in the early 1970s. Flow cytometry was developed in the 1950s, and provided the basis for automated hematology analyzers introduced a decade later. These technological advances ensconced the clinical laboratory as a routine and essential component of medical care because laboratory tests were not just available, suddenly they were abundant and rapidly available. What is more, the vast majority of laboratory tests were reimbursed by Medicare, Medicaid, or private insurers. The economics of clinical laboratory services were weighted towards expanding laboratory facilities, staff, and utilization, and there was a proliferation of local, regional, and national medical laboratories.

Spiraling costs of medical care forced changes in reimbursement strategies in the 1980s. Two innovations were introduced:

- Prospective payment based on diagnosis-related groups (DRGs), in which payment for healthcare was determined by the diagnosis rather than the specific services rendered.
- Health Maintenance Organizations paid a flat fee per enrolled member to healthcare providers, and these contracts were sometimes referred to as "full risk."

These reimbursement models had a dramatic effect on the financial impact services such as laboratory, radiology, and pharmacy had on hospital budgets. In a fee-for-service model, these departments were revenue centers for hospitals because each individual service generated revenue. However, prospective payment and full-risk contracts created a financial environment in which these services were costs measured against predetermined reimbursement and laboratories were forced to minimize both the cost and the utilization of their services.

Nowadays, healthcare reimbursement is a mixture of prospective payment, full-risk, and fee-for-service models, and facilities attempt to balance and maximize these sources of revenue. Fee-for-service is the most financially attractive model for healthcare providers, but is mostly limited to private insurers. Many hospitals serve populations that include significant numbers of uninsured patients and those covered by full-risk contracts mandated by local governments for medical care of constituents unable to afford health insurance. An unfavorable mix of fully insured, full-risk, and uninsured patients puts great pressure on healthcare facilities to minimize their costs.

A 2007 Washington G-2 Report cited by the Centers for Disease Control and Prevention (Terry, 2007) estimated that laboratory tests account for only 2.3 % of healthcare expenditures in the USA, yet laboratory budgets often are the target of cost-cutting initiatives. The following discussion will focus on the budgetary strategies that can minimize the operational costs of clinical laboratories.

Personnel

As with the rest of the healthcare system, personnel costs are the largest component of a laboratory budget, and consequently receive the greatest scrutiny when laboratory budgets need to be cut. Qualified staff is a requirement for laboratory accreditation, and several states license medical laboratory personnel based on training, experience, and certification by state or national examinations. CLIA '88 established minimum educational requirements for laboratory personnel performing waived, moderately complex, or highly complex procedures. These requirements drive the portion of laboratory budgets devoted to personnel costs. Under federal law, waived tests can be performed by personnel with a high school education as long as competency requirements are met. Highly complex tests require technical personnel with at least an associate's degree or 60 h of college credit with minimum requirements in laboratory sciences or a degree in medical laboratory technology from an accredited institution. The education and experience requirements under CLIA '88 for personnel performing moderately complex tests are considerably more flexible, ranging from an associate's degree to a high school diploma with documented experience, training, and competency.

In 2003, the CLIA law was revised, and the classification of laboratory tests was simplified to "waived" and "non-waived," effectively eliminating the distinction between moderately and highly complex tests; the latter category now applies mostly to tests that have not been FDA approved, or have been modified for uses not included in the 510K application approved by the FDA.[2]

Some states require laboratory technical personnel to be licensed, and the requirements for licensure may exceed the federal standards. Agencies that accredit medical laboratories also apply minimum qualifications for technical personnel that may go beyond the federal standards. The balance between technical (laboratory staff who report results) and nontechnical (support staff assigned to specimen processing, phlebotomy, and clerical functions) is an important consideration that will greatly impact the laboratory budget since technical personnel command salaries that are twice or more the salaries of nontechnical staff. Recent data place the median salary of medical technologists certified by the American Society for Clinical Pathology (ASCP) at nearly $60,000.[3] Medical laboratory technicians (MLTs), for which certification is offered but educational and training requirements are not as demanding as for medical technologists, typically earn around $30,000. Uncertified laboratory personnel usually earn less than MLTs.

Partly due to economics, and partly due to a shortage of qualified technologists, there has been a gradual shift towards lesser qualified technical personnel in clinical laboratories since the 1980s. The newer requirements under CLIA '88 classified most laboratory tests as moderately complex (and later, simply "non-waived"), which could be performed by staff with minimal training. Most laboratories still employ certified medical technologists, but often have MLTs on their staff, as well. Staff who function in supervisory or managerial capacities are required to meet higher standards of education and training.

---

[2] In the pharmaceutical industry deviations from FDA-approved indications is termed "off-label," but in laboratory practice, it generally means any modification of the manufacturer's specifications for performing an approved test. The FDA approves laboratory tests for in vitro diagnostic use based on validation data in the manufacturer's 510K application. The diagnostic interpretation of laboratory tests, and their appropriate use, is not restricted. Only the way the test is performed is subject to FDA restrictions. Use of non-FDA approved tests by clinical laboratories currently is a controversial issue, focusing primarily on molecular diagnostics. "Lab-Developed Tests," or LDTs, are the focus of intense scrutiny by regulatory agencies, and a consistent set of criteria for their validation has not yet been developed.

[3] The ASCP is one of several organizations that certify medical laboratory professionals, but it is the oldest and most widely recognized certifying agency. The American Board of Bioanalysis (ABB) and American Medical Technologists (AMT) also certify medical technologists.

Many laboratories engage PRN[4] (part time, as needed) technologists so they can be flexible with staffing. Part-time employees are less costly to the institution since they ordinarily do not receive benefits, and are used only when work volume is high enough to need them. Full-time employees inevitably experience certain times when the workload is low and the laboratory is, for a period of time, overstaffed. Using PRN employees allows laboratory administration to increase and decrease staff within a relatively short period of time in response to changes in the workload.

Analytical Platforms

In the automated areas of a clinical laboratory (chemistry, hematology, and perhaps urinalysis), the proper choice of analyzers (the "analytical platform") is critical to the long-term success of the laboratory operation. Choosing a platform with sufficient throughput to dispatch specimens quickly even during the periods of highest work volume is one of the most important considerations. An analyzer that is only just able to generate test results sufficient to keep up with the average pace at which specimens are received will rapidly be inundated if there is a sudden spike in the number of specimens, or if unanticipated maintenance is required on the instrument, allowing specimens to accumulate. Other than personnel, the instrument and reagent costs of performing the tests are the largest expense for clinical laboratories, and a poor choice of an analytical platform can result in greater than necessary expenses, poor turnaround times, greater demands on staffing, and substandard quality of the laboratory services.

There is a balance between throughput and cost of analytical platforms. Low throughput instruments are typically less expensive to purchase, but often have a higher operating cost per unit because reagents are packaged in smaller quantities and many low-volume instruments require greater use of disposable components compared to instruments designed for higher throughput. On the other hand, high-volume instruments are considerably more expensive to purchase, but due to economies of scale the reagent cost per unit is often minimized. For this reason, often it is more economical in the long run to purchase an analytical platform that has greater throughput than the work volume requires because of lower operating costs.

There are three principal approaches for financing automated chemistry or hematology analyzers:

Capital purchase: In this approach, the equipment is purchased with capital funds, much the same as purchasing an automobile with cash rather than a loan.

Lease/purchase or reagent rental: In a lease/purchase agreement, a monthly lease payment is paid for the equipment, which is amortized over, typically, 5–7 years. This would be equivalent to leasing an automobile, and some of these agreements have a buyout clause that allows purchase of the equipment for its residual value at

[4] From the Latin *pro re nata*, meaning "in the circumstances" or "as the circumstance arises."

the end of the lease. Reagent rental agreements are similar to a lease, except the laboratory agrees to purchase a minimum amount of reagents, and the cost of the equipment is built into the cost of reagents. To some extent, the distinction between these two approaches is only important because of where the costs appear in the budget and get accounted for by the institution. In both cases, the vendor is placing the equipment in the laboratory for an agreed upon price, which is spread out over several years in either lease payments or reagent surcharges.

Cost per test: In this type of agreement, the laboratory pays for neither the instrument nor the reagents, but instead pays the vendor for every result that is reported. The equipment is wholly owned by the vendor. Unlike most lease/purchase agreements, in the cost per test contract the laboratory builds no equity in the equipment over time. All reagents and supplies are provided to the vendor, and the laboratory pays a fee to the vendor for each test result reported.

These three models represent the spectrum from least overall cost to most costly, and correspondingly the most risk to least risk. By far the most economical way to purchase equipment is with capital funds. Any other method entails some form of financing that inevitably will inflate the price. Outright purchase of equipment does not require a contractual obligation to purchase reagents from the same vendor, and alternate reagents from competing vendors are sometimes compatible with the platform and available at a lower price. By purchasing the equipment, the laboratory retains bargaining leverage, both in the price of the instrument and the reagents to operate it.

Along with the cost advantage, however, capital purchase of laboratory equipment involves the greatest risk to the laboratory because once the platform is purchased it becomes a depreciable asset[5] of the organization over the term of its useful life—typically 5 or 10 years, depending on the total value of the asset. If the laboratory's needs for an analytical platform change within that period due to, for example, expansion or reduction of services, the institution is left with few options for replacing the platform without having to write off the residual value of the equipment. Occasionally, the analytical platform does not perform to expectations and becomes a drain on laboratory resources, sometimes even compromising the quality of laboratory services. These are the risks of capital purchases. As when purchasing an automobile with cash, where the owner accepts all the risk in the event the vehicle is a lemon, rectifying the bad investment will involve a substantial loss. Therefore, a laboratory should have complete confidence that the analytical platform will meet their needs over the depreciated lifetime of the asset before making the investment. But when the equipment performs as expected, and meets the needs of the laboratory and institution over its useful lifetime, the capital investment offers the greatest financial advantage.

A fully automated chemistry or hematology platform for a clinical laboratory serving a moderately sized hospital (e.g., approximately 600 beds) is likely to cost

[5] Depreciation is an accounting mechanism for businesses to deduct the decrease in value of a capital asset from their profits, thereby recovering some of the cost of their investment. US tax laws allow this deduction to encourage businesses to invest their capital in assets that enhance productivity.

$1–2 million, and many hospitals do not have sufficient capital reserves to make that kind of investment (just like many people don't have enough savings to pay cash for an automobile). In circumstances where capital funds are not available payment has to be spread out over several years. Lease and reagent rental agreements provide alternatives that do not require capital but instead shift the cost of the equipment into the operating budget, in payments spread out over the useful life of the platform.

Lease and reagent rental agreements represent shared risk among the vendor and purchaser. These contracts can be complex, involving many performance standards for the analytical platform and rights to terminate the agreement with notice. The leverage retained by the laboratory is that payments can be discontinued without loss of the entire value of the equipment if the terms of the contract are not satisfied. The contracts often contain language that provides for upgrades when new technology is available that may benefit the laboratory operation. A good analogy is renting an apartment: if the landlord does not maintain the property in a manner guaranteed in the lease, the tenant has the option of breaking the lease without significant financial loss, other than the unanticipated expense of moving to another apartment and costs associated with entering into another lease.

For the flexibility of changing analytical platforms as the laboratory needs evolve, or because the platform does not perform to expectations, the laboratory accepts a higher cost of both the equipment and the reagents it uses. The difference between lease and reagent rental agreements is largely superficial; in either case the vendor is charging a financing fee for the use of the equipment. In a lease agreement, the total of the lease payments will exceed the price for which the equipment would be offered in a capital purchase. In a reagent rental agreement, the lease payments are added as a surcharge to the cost of supplies and reagents, with a minimum sales volume stipulated in the contract to ensure the vendor is compensated for the cost of financing the equipment. These agreements result in higher operational costs for the laboratory, but no capital investment has been made, so the financial risk to the institution, and the operational risk to the laboratory in the event the analyzer does not meet its needs, is less than if the analytical platform had been purchased. The vendor accepts the risk that the installed equipment may not perform to negotiated specifications, resulting in loss of the contract. In addition, capital the vendor has invested in the equipment is paid back slowly over time, and is not available for immediate reinvestment.

In a strictly cost per reportable result agreement, the vendor provides the equipment, supplies, and reagents, accepting the risk that the test volume will generate adequate income to compensate them for the value of the instrument. The laboratory has little risk in these agreements, since it has not expended capital and relies on reimbursement for the laboratory tests to offset the cost paid to the vendor when a chargeable result is generated. For laboratories that operate in healthcare environments where reimbursement for laboratory tests is mostly guaranteed, these contracts are attractive. However, laboratories that serve patients insured under full-risk or prospective payment models incur high costs for laboratory tests under cost per reportable agreements since the cost to the laboratory for each reported test is maximized to include equipment amortization and the overhead accepted by the vendor for providing all of the resources necessary to do the test.

Generally, cost per reportable result contracts are not an attractive option for central laboratory services. These arrangements are most suited to physician office and satellite laboratories where space and personnel limit the choices of analytical instruments that can be used.

A final consideration in the cost of an analytical platform is service. A clinical laboratory that serves a hospital offering emergency and acute care services must have testing available at all times, and therefore technical problems with analytical equipment have to be resolved as quickly as possible. Newly purchased equipment ordinarily is covered under a manufacturer's warranty for at least a year, and the warranty should ensure that service personnel will respond promptly to any problems with the equipment. Beyond the warranty period, the laboratory will have to pay for a service agreement that ensures the same response to service issues. The price of the service contract usually increases as the instrument gets older, since more frequent service is expected. Lease and reagent rental contracts typically include the cost of the service agreement in the annual minimum specified in the contract. Cost per reportable test arrangements usually include any service required.

The Laboratory Budget

A laboratory operating budget includes two primary components: fixed costs, which are independent of test volume, and variable costs, which are determined by the number of tests performed. Not all expenses fit neatly into one or the other of those categories; some costs are partly fixed and partly variable.

Examples of fixed costs are:

- Building and equipment amortization
- Service contracts on equipment
- Utilities
- Accreditation
- Facility maintenance
- Licensure (where applicable)

All of these costs are influenced, to some degree, by the volume of work produced in the laboratory. A small laboratory, for example, has lower utility and maintenance costs than a large laboratory. Accrediting agencies usually adjust their fees based on the work volume. Small, low-volume laboratories have lower building and equipment costs than larger laboratories. However, within the environment a particular laboratory operates, these expenses will not vary substantially from month to month or year to year, even if the laboratory experiences fluctuations in work volume. There are strategies to reduce these fixed costs, such as energy efficient buildings and limiting equipment to the minimum necessary to produce the work required of the laboratory, but personnel and variable expenses provide the most fruitful opportunities for minimizing costs.

Personnel costs, the largest portion of any laboratory budget, are partly fixed and partly variable. Salaries for administrative and managerial personnel are mostly fixed, and are determined by the institutional and accrediting agency requirements for oversight of laboratory services. Although in general larger laboratories require a larger administrative and managerial staff, within the broad categories of clinical laboratory services—e.g., hospital, referral, physician office—certain positions are necessary to ensure the laboratory has sufficient supervision. The same is true for some nontechnical support personnel, such as clerical, phlebotomy, and specimen processing personnel. Some staffing adjustments can be made when work volume increases or decreases, but for the most part salaries for the nontechnical, administrative, and managerial staff are fixed expenses in the laboratory budget.

Salaries for technical staff are the most variable among personnel expenses. Clinical laboratories closely monitor work volume to ensure that staffing is appropriate. When work volume increases significantly, additional technical staff may be required, and the opposite is true when work volume decreases. These adjustments ordinarily are made during the annual budget cycle, although staffing adjustments in mid-year can be precipitated by significant changes in the laboratory work volume such as adding a new, large, outreach client or the closure of a hospital service that previously generated a large volume of laboratory work. The variable component of personnel costs is not incremental, as are the strictly variable costs discussed below, but instead is stepwise as staff positions are added or eliminated in response to changes in the workload. It was mentioned earlier that PRN staff are an efficient way to adjust personnel to short-term changes in workload, but it is difficult to budget PRN wages since most times, the changes in workload cannot be predicted. In practice, PRN staff are mostly used to cover for full-time salaried staff on vacation or sick leave.

Variable laboratory costs include:

- Reagents to perform tests
- Disposable supplies such as collection tubes, transfer pipettes, labels, etc.
- Distilled water, bottled gas
- Forms, printer paper, toner

Next to personnel costs, reagents usually are the largest component of a clinical laboratory budget and vary from a few cents to over $100 per test. Managing reagent costs is one of the most important strategies for operating cost-effective laboratory services. Reagent costs are affected by the instrumental platforms chosen and the mechanisms by which the analytical instruments are financed, as discussed in the previous section. Capital purchase of the analytical platform minimizes reagent costs, whereas low-risk cost per test contracts are the most expensive. Often, a central laboratory service has a mixture of purchased and leased equipment, and perhaps some low-volume platforms that are used on a cost per test basis.

Decades ago it was common for clinical laboratories to make their own reagents, either because commercial products weren't available or because it was cheaper to make their own reagents than to purchase the premixed reagents from a vendor.[6] That approach is no longer practical since diagnostic reagents are subject to extensive validation requirements, and most laboratory methods use reagents approved by the FDA. Prepackaged diagnostic reagents are standard, and comprise an essential component of laboratory budgets.

The reagent budget for a clinical laboratory is not entirely variable because quality control (QC), calibration, and proficiency challenges consume reagents and are required to make patient tests available, but are not related to the number of patient tests performed. For all but cost per test contracts, the expense of reagents to periodically calibrate the method, perform QC, and report proficiency assessments to satisfy accreditation standards represent unreimbursed overhead that is mostly independent of workload. Low-volume tests that are not urgent can be scheduled in a way that minimizes the overhead, but some tests need to be available at all times and there are circumstances when the cost of calibrating and running QC exceeds the cost of analyzing patient specimens. For high-volume tests the impact of overhead costs are minimized as a proportion of the overall cost of providing the service.[7]

Other factors influence the unit cost of reagents. High-volume accounts usually receive favorable pricing, and discounts often cross platforms (i.e., when reagents are purchased from a single vendor for multiple tests and platforms). There is considerable incentive for the vendors of laboratory equipment and diagnostic reagents to pursue contracts that establish them as the principal source of multiple laboratory platforms for a healthcare facility, and consolidation of reagent purchase contracts with a single vendor gives leverage to the laboratory for negotiating more favorable terms.

Alignment with a purchasing consortium is another way to take advantage of volume to get favorable pricing on analytical reagents. Vendors of clinical laboratory diagnostic equipment and reagents negotiate contracts with consortia for discounted prices that are commensurate with the scope of the overall agreement; the larger the consortium, the more favorable the terms of reagent contracts.

---

[6] Those in the older generation of laboratory professionals may recall when the DuPont ACA (automated clinical analyzer) was introduced and required the use of vendor-provided distilled water; use of any other purified water was not supported under the terms of the service agreement. This caused some indignation among laboratory directors, since all laboratories had a supply of distilled water, and paying for it as a reagent seemed unnecessary. It was a harbinger of things to come, however. Laboratory reagents nowadays are sold as a complete package, and modification of any component is not allowable under the terms of FDA approval.

[7] This is a generalization that ordinarily holds true, but exceptions exist when a test requires frequent calibration and QC because the method is unstable. Overhead can increase when the test volume increases if, for example, QC is required every few specimens to ensure analytical drift has not affected results. But that circumstance is unusual, and mostly limited to manual tests that involve errors associated with technique. An example is a chromatographic method that requires controls in parallel with every patient specimen.

Laboratory Design

Each subspecialty in laboratory medicine requires unique skills. Transfusion medicine services perform relatively few laboratory tests, but are responsible for issuing blood products for therapeutic use. Microbiology laboratories involve mostly manual diagnostic methods to identify pathogens in blood, urine, or other specimens. Hematology, chemistry, and immunology are highly automated and represent the largest volume of routine laboratory tests. Molecular diagnostics and flow cytometry involve automation, but these methods usually require specialized training of the technical personnel and interpretation by a qualified pathologist or molecular biochemist.

Services and Laboratory Consolidation

Historically, the design of clinical laboratories segregated the various services, but that approach involved the highest personnel costs; each laboratory needed sufficient technical and supervisory personnel to ensure both compliance with accreditation standards and that services were available at all times. Modern laboratory designs consolidate services to minimize space and the number of technical and supervisory personnel required. In smaller laboratories, technical personnel may perform tests in multiple laboratory areas—e.g., microbiology, transfusion services, hematology, and chemistry. In larger laboratories, however, this approach is not practical because the technical expertise required in each area of a clinical laboratory that offers extensive services is too demanding to expect technologists to remain proficient in more than one or two areas. The core laboratory concept, which combines the highly automated areas of chemistry and hematology, emerged in the 1980s and dominates the clinical laboratory landscape today.

Grouping automated laboratory procedures into a core laboratory had several advantages. First, it simplified the logistics of specimen distribution in the laboratory since the highest volume procedures were concentrated in one area. Second, it allowed essential resources such as centrifuges, printers, and workstations to be shared between chemistry and hematology, avoiding duplication of those resources. Finally, the core laboratory encouraged cross-training of chemistry and hematology technologists, which allowed more flexibility and efficiency in scheduling staff. Since both specialties depended primarily on automated platforms, chemistry and hematology shared many aspects of laboratory practice, and therefore were easily combined.

Microbiology is the highest volume area of the laboratory after the core laboratory (chemistry and hematology), but is unique in several respects. The equipment required for microbiology, such as biohazard hoods, incubators, and microscopes, is not widely used in other laboratory areas (with the exception of microscopes in hematology). Most of the microbiology procedures are manual[8] and involve skills that do not translate easily into a core laboratory environment, which focuses primarily on high-volume testing.

Transfusion services are unique because in addition to performing laboratory tests, they also issue blood products for therapeutic use. The number of laboratory procedures performed by transfusion services is limited, and most are manual. When their function includes modification or relabeling of blood products, there is an additional layer of regulatory oversight; in addition to compliance with CLIA requirements, these laboratory services must comply with FDA regulations. The FDA regularly inspects blood collection centers to ensure compliance with standards recommended by the Blood Products Advisory Committee. Transfusion services that neither collect their own blood products nor modify or relabel products they purchase and issue are not regularly inspected by the FDA, although they are subject to unannounced inspections. Like microbiology, the workflow and technical skills characteristic of transfusion services are mostly incompatible with core laboratory designs, so these services are often segregated from other parts of the laboratory.

Two laboratory services have characteristics that are intermediate between the highly automated core laboratory and manual test-oriented microbiology and transfusion services: serology/immunology and urinalysis. Most serological tests are automated on immunoassay platforms, and therefore may be incorporated into the chemistry services, which also include immunoassays. Many chemistry/immunoassay platforms have serological tests available, and performing these tests in the core laboratory provides all the advantages of consolidation described above. However, states that license medical technologists typically do so only in the specialties in which the licensee is qualified. Graduates of accredited medical technology schools who have passed national certifying exams usually are licensed in all specialties,[9] but continuing education in each specialty may be required to maintain licensure in all areas. As a result, some laboratory staff maintain only one or two specialties on their license, to reduce the burden of meeting the CE requirement. This presents a problem if, for example, a technologist with a license only in chemistry is performing serology tests on the automated chemistry platform, since a license in serology may be required by the state for the technologist to perform those tests. This is only a concern in states that regulate clinical laboratories and license laboratory personnel.

Urinalysis might logically be located in chemistry, hematology, or microbiology. Clinical laboratories that process many urine specimens typically use an automated

[8] Automation of some microbiology procedures may be on the horizon. There is growing interest in using matrix-assisted laser desorption ionization/time-of-flight (MALDI-TOF) mass spectrometry to identify microbes by their protein signatures (Lay, 2001).

[9] The terminology for medical technology training programs is changing, and many programs now are called "Medical Laboratory Science" or "Clinical Laboratory Science" programs. The terminology used by certifying agencies also has changed. The credentials awarded by the ASCP to those who pass their medical technologist exam used to be MT(ASCP), but this was changed to MLS(ASCP) when the ASCP Board of Registry merged with the National Credentialing Agency (NCA) in 2009.

urinalysis platform, and in those laboratories urinalysis is most conveniently located in the core laboratory. Manual urinalyses might fit better in microbiology laboratory because of the microscopy component.

Molecular diagnostics are still evolving from their beginnings as primarily a specialty service operated by laboratory professionals trained in the emerging science of molecular biology. Molecular methods now are in the mainstream of clinical laboratory services, and the instrument platforms available for these assays are progressing towards automation. At some point, molecular methods may fit into core laboratory services as another automated platform, but that point has not yet been reached. Currently, molecular diagnostics laboratory services usually have dedicated staff and may be located in facilities remote to the main laboratory. With regard to diagnostic use, there is substantial overlap between molecular pathology and other clinical laboratory services. Molecular methods may be used to amplify and characterize DNA from infectious agents, complementing microbiology services; identify genetic mutations associated with tumors, complementing biochemical markers of neoplasia; and to reveal pharmacogenetic polymorphisms to optimize drug therapy, complementing therapeutic drug monitoring services.

Support Services

Laboratory support services can include phlebotomists, specimen processing clerks, laboratory assistants, phone operators, and other personnel not involved in the technical component of laboratory testing. Support personnel usually are needed in all but the smallest of laboratories to receive, label, centrifuge, and distribute specimens to the analytical areas. In addition, support personnel often are responsible for answering phone calls to the laboratory, preparing specimens to be sent to other laboratories for referral tests, delivering laboratory reports when electronic delivery is not available, restocking laboratory supplies, and various other nontechnical duties.

In the design of a clinical laboratory, a common error is failure to allocate sufficient space for specimen processing. The more crowded the specimen processing area, the greater the likelihood of specimens being misplaced or overlooked. There should be generous space to sort and organize specimens arriving in the laboratory. When there is insufficient space for efficient specimen processing, it often becomes a bottleneck in the overall workflow of the laboratory. The advantage of high-throughput automated analyzers is lost if significant delays occur between the time a specimen is received and when it gets distributed to the analytical area. A goal of laboratory design should be to minimize the time required for delivery of specimens to technical personnel because they are the highest paid employees working in the laboratory; technologists' time that is wasted when they wait for specimens already in the lab is more expensive than, for example, time spent by nontechnical personnel waiting for specimens to be delivered to the lab. Another way to state this principle is that productivity of the highest paid employees should be maximized at the expense of less productivity in lower paid staff, if necessary. Achieving this goal may mean adding another specimen processing clerk to the staff, which may give them all less to do, but ensures that no delays occur in getting specimens to the technical personnel.

Not all laboratories provide phlebotomy services. Referral laboratories almost always establish phlebotomy stations strategically located throughout their market. Hospital laboratories may provide phlebotomists, or the hospital may depend on nurses and medical assistants to collect blood specimens. Phlebotomists are skilled professionals,[10] although few states require licensure of phlebotomists. In hospitals, whether blood collection is performed by laboratory employees, patient care staff, or some combination of the two is mostly an institutional decision. There are several advantages to having phlebotomists part of the laboratory staff:

- The laboratory trains the phlebotomists and monitors their competency.
- The logistics for deployment of the phlebotomists is determined with laboratory workflow in mind.
- Changes in phlebotomy procedures can be rapidly implemented.
- The laboratory can control the utilization of phlebotomy supplies.

The advantages of having phlebotomists as part of the laboratory staff are difficult to overstate. Within the laboratory services, phlebotomists' training will be consistent and address the principal quality issues associated with blood collection: proper patient identification, collection of a sufficient volume of blood in the appropriate container, phlebotomy techniques that minimize the risk of hemolysis, and proper labeling of the specimen to avoid the need for relabeling when the specimen reaches the laboratory. Specimens that need to be relabeled in the laboratory, and particularly specimens that cannot be analyzed due to improper collection, have enormous impact both on the efficiency of workflow in the lab and on good patient care, since re-collection causes unnecessary discomfort and risk to patients. In addition, mislabeled specimens present a risk of misdiagnosis and improper treatment.[11] Rigorous, consistently enforced patient identification procedures minimize this risk.

A phlebotomy staff has significant impact on the laboratory budget, however, and it is tempting for laboratory administration to transfer those costs to other departments

[10] There are several national organizations that certify phlebotomists, including the National Healthcare Association, National Center for Competency Testing, and National Phlebotomy Association.

[11] In reference to laboratory specimens, "mislabeled" can have several meanings, including labels that do not have sufficient information, or barcode labels that cannot be scanned because they are not properly affixed to the collection tube. In this context, however, "mislabeled" refers to a specimen labeled with information identifying a patient different than the one from whom blood was collected. A few studies have estimated the frequency of mislabeled ("wrong blood in tube" or WBIT) specimens received in hospital clinical laboratories. These studies are difficult to design and conduct because without sophisticated (and expensive) genetic analysis, it often is not possible to determine whether the blood in a collection tube belongs to the patient whose name is on the label. Studies have consistently identified specimen mislabeling as a common source of error in laboratory results, and estimate that the frequency of WBIT is between 0.05 and 0.1 % (Ansari & Szallasi, 2011; Wagar, Tamashiro, Yasin, Hilborne, & Bruckner, 2006).

such as nursing or other patient care services in an effort to reduce laboratory personnel costs and thereby improve productivity metrics that are based on revenue per employee. Efficiency of workflow, minimizing unnecessary re-collections, and improving patient care by reducing risks associated with mislabeled specimens are difficult to quantify and are not usually reflected in simple calculations based on total revenue and personnel costs in a laboratory budget. However, these factors should be given serious consideration when deciding how best to provide phlebotomy services in hospital settings.

Laboratory Automation

Several manufacturers of clinical laboratory instruments offer automated systems that are capable of performing certain specimen processing tasks prior to, and after, analysis. The sophistication of these systems ranges from specimen input stations that serve multiple analyzers and direct specimens based on information contained in the bar-coded label, to fully automated systems (total laboratory automation, TLA) that process specimens from the time they are delivered to the laboratory to archived storage following analysis. TLA systems represent the extreme in a spectrum of automated specimen processing systems, where a specimen arriving in the laboratory is placed on an automated track that reads bar-coded information on its label including the patient ID and the tests to be performed; the specimens are routed to a processing station that may include centrifugation and transfer of aliquots to additional test tubes, the processed specimens are directed to a track leading the appropriate automated analyzer, the tests are performed, and the remaining specimen is transported along another track to an archiving facility that logs its location so it can be retrieved for subsequent laboratory requests, if necessary. Technologists monitor the laboratory results and intervene only when required, such as when results are critical or otherwise needing attention. Most laboratories that use automated specimen processing systems do not have a TLA-type system, but have some degree of automation that eliminates the need to perform certain specimen processing steps manually.

The principal advantage of automated specimen processing systems is they perform tasks that otherwise would require personnel. Although automated systems are expensive to purchase and service, laboratories with sufficient workload to justify automation can achieve significant savings due to reduction in the number of support staff needed to process specimens. As an example, if an automation system eliminates the need for six support personnel (two per shift), each earning $25,000 annually, the savings over a 10 year lifetime of the system is $1.5 million, which is sufficient to offset the cost of all but the most expensive systems.

A second advantage of automated specimen processing is consistency. When specimens are processed manually, preanalytical variables are introduced that may have an effect on the analytical results. These variables include the time between receiving the specimen and performing the analysis, centrifugation speed and time, pour-off technique and volume, and whether specimen tubes are maintained upright or horizontal (i.e., lying on a bench top). Automated systems treat all specimens the same way, reducing the chance that preanalytical factors will cause variability in analytical results. Automated systems are not necessarily faster than manual processing, but are generally more reliable and consistent.

Finally, automated specimen processing systems offer the advantage of better control of the workflow because the location of a specimen in the laboratory is traceable at any time. This is a useful feature when, for example, the laboratory receives an inquiry about the status of a particular test that has been ordered. Without automation, laboratory personnel may have to leave their workstation to search for the specimen, whereas automated systems track the location of every specimen in the laboratory, and the information is easily retrievable. Automated systems also generate data that can be used for quality assurance and performance improvement initiatives. For every progression a specimen makes through the automated system a time stamp is recorded, so in-laboratory turnaround time data can easily be retrieved and monitored.

For laboratories that have sufficient work volume to justify it, some degree of automated specimen processing technology usually is a good investment that reduces the number of nontechnical staff required and improves the overall laboratory performance.

Point of Care Testing

Certain types of laboratory tests are deregulated if they meet the criteria set forth in CLIA '88 (Subpart A Section 493.15 F.C.):

- Are cleared by FDA for home use
- Employ methodologies that are so simple and accurate as to render the likelihood of erroneous results negligible or
- Pose no reasonable risk of harm to the patient if the test is performed incorrectly

If a laboratory test meets these criteria, the FDA may grant it a "waiver," which effectively exempts the test from regulation under CLIA.[12] Waived tests proliferated under these new rules, and now comprise a large segment of laboratory testing in the USA.

Because they often are used outside of a central laboratory, waived tests have been variously called "near patient" or "point of care" (POC) tests; the latter term is more common. POC tests include blood glucose, urine hCG to detect pregnancy,

[12] Healthcare facilities, including private physicians' offices, are required to obtain a Certificate of Waiver from CMS before performing waived tests, but the tests are otherwise unregulated by the government. However, accrediting agencies such as The Joint Commission and CAP apply certain standards to waived tests. These accreditation standards for waived tests include specifications for training and periodic verification of competency, QC, documentation of results and reference ranges, and needs assessment.

urinalysis by solid reagent (dipstick) technology, fecal occult blood, drugs of abuse, and a few other tests that have been granted a waiver by the FDA. Not all laboratory tests performed outside the central laboratory are waived; there are no federal restrictions on where laboratory tests can be performed, so any test can be performed at the point of care as long as the regulatory and accreditation standards are met. Hence, POC does not necessarily refer to waived tests, but the vast majority of tests performed at the POC are waived. Exceptions include several non-waived platforms that are portable (either on a cart or hand-held) and perform blood gas, electrolyte, cardiac marker, lactate, and a few other tests.

Non-waived laboratory tests performed outside of a laboratory nearly always involve higher direct costs, and are subject to the same regulatory and accreditation standards, as tests performed in a laboratory. This creates significant overhead costs. Therefore, implementing non-waived POC tests in a healthcare facility that has a central laboratory offering the same tests at lower cost requires careful consideration of the quantifiable benefits that justify performing the tests at greater cost at the POC. Two settings where non-waived POC tests are particularly useful are the emergency department (ED) and operating room (OR). In the ED, rapid results are helpful in triaging patients to appropriate areas, which improves efficiency of emergency medical treatment. In the OR, rapid assessment of coagulation, blood gas, and electrolyte status is essential for keeping anesthetized patients stable during surgery. In both of these settings, a core laboratory has difficulty providing the turnaround time that is either beneficial (ED) or essential (OR) for proper medical care. Implementation of non-waived POC tests in other healthcare settings, when central laboratory services are available, is more difficult to justify, since it is more of a convenience to healthcare providers than an overall improvement in patient care.

There is one advantage of POC testing, however, in virtually all healthcare settings. The instruments designed for non-waived POC tests use very small volumes of blood compared to collection tubes for laboratory tests performed on automated platforms. This is a particular advantage with pediatric patients.

Most POC tests are waived, so the regulatory and accreditation requirements are less than for non-waived tests. Although many of the waived POC tests have higher direct costs compared to the same tests performed in the lab, the difference is not as large as for non-waived tests. Hence, waived POC testing is very common in healthcare settings. Blood glucose is by far the most common POC test performed on hospital inpatients, since there have been studies indicating that prevention of hyperglycemia in hospitalized patients improves outcomes (Van den Berghe et al., 2001). Although the value of glycemic control in hospitalized patients has been questioned (Boyd & Bruns, 2001; Wiener, Wiener, & Larson, 2008), currently it is considered the standard of care.

A caveat associated with waived POC tests is that the analytical performance standards to which they are held during the FDA approval process are far less than standards for non-waived tests. Waived devices for measuring blood glucose, for example, only need to meet a ±20 % accuracy standard, whereas most clinical laboratory glucose methods are accurate within 1–2 %. Clinicians may not understand that difference, and expect POC glucose results to match laboratory results, when often there will be considerable bias between the two. Another misconception with regard to glucose devices is that certain waived blood glucose meters have been approved "for professional use," and presumably are more accurate than "home use" devices that diabetic patients use to monitor their own blood glucose. The FDA does not have dual standards for waived blood glucose devices; all waived glucose devices are merely required to meet the ±20 % standard for accuracy. Manufacturers may label these devices as "home" or "professional" use, but the distinction is for marketing purposes, and does not necessarily reflect any difference in analytical performance. "Professional" devices usually have connectivity capabilities that make it possible to capture data in the laboratory or hospital information system.

Since waived tests, by definition, are subject to minimal regulation, administrative oversight of a waived POC testing program does not require the input of laboratory professionals, and at one time many institutions managed POC testing independently from the laboratory. However, about a decade ago the accreditation standards for waived testing became stricter, and the input and participation of laboratory professionals in managing waived POC testing became more important. Laboratory directors, managers, and supervisors are accustomed to the process of laboratory accreditation and are better suited than non-laboratory personnel for ensuring that waived testing complies with accreditation standards.

Summary

Laboratory tests have been a part of medical care at least as long ago as the time when Hippocrates practiced, around 300 BC. The first clinical laboratories were established late in the nineteenth century. The results of clinical laboratory tests are used to diagnose disease, determine appropriate therapies, detect toxins, monitor therapeutic drug concentrations, and assess overall health. Laboratory tests consume a very small fraction of total healthcare spending, but have great influence over medical decisions. There is little doubt that laboratory services are essential to adequate healthcare, and their role is certain to increase as clinical applications of new technologies such as proteomics and molecular diagnostics expand.

Clinical laboratory services are provided in a variety of settings: physicians' offices, clinics, hospitals, and regional and national referral centers. Each of these settings has unique requirements for diagnostic laboratory services; therefore the design of laboratory services depends highly on the environment in which it operates. Virtually every consideration in clinical laboratory design and function—personnel, equipment, automation, consolidation, test menu, location, etc.—is influenced by the type of services the laboratory needs to provide. The greatest challenges in clinical laboratory design and function are faced by hospital laboratories because they serve the most diverse set of laboratory needs: inpatient, outpatient, critical, routine, referral, and outreach.

A well-designed laboratory service is an asset to the institution it serves, providing timely laboratory results that are used to improve patient care. In addition, prudent choices for equipment and personnel, and attention to designing efficient processes that do not waste time or resources, benefit the institution by producing an essential service at a competitive cost. In some circumstances, a greater investment results in long-term savings. Also, assigning laboratory-related services such as phlebotomy and POC testing to supervision by laboratory staff has potential benefits for both the laboratory and institution.

References

Ansari, S., & Szallasi, A. (2011). 'Wrong blood in tube': Solutions for a persistent problem. *Vox Sanguinis, 100*, 298–302.

Boyd, J. C., & Bruns, D. E. (2001). Quality specifications for glucose meters: Assessment by simulation modeling of errors in insulin dose. *Clinical Chemistry, 47*, 209–214.

Lay, J. O., Jr. (2001). MALDI-TOF mass spectrometry of bacteria. *Mass Spectrometry Reviews, 20*, 172–194.

Plebani, M. (2006). Errors in clinical laboratories or errors in laboratory medicine? *Clinical Chemistry and Laboratory Medicine, 44*, 750–759.

Terry, M. (2007). *Lab industry strategic outlook: Market trends and analysis*. Washington G-2 reports. New York, NY.

Van den Berghe, G., Wouters, P., Weekers, F., Verwaest, C., Bruyninckx, F., Schetz, M., et al. (2001). Intensive insulin therapy in critically ill patients. *The New England Journal of Medicine, 345*, 1359–1367.

Wagar, E. A., Tamashiro, L., Yasin, B., Hilborne, L., & Bruckner, D. A. (2006). Patient safety in the clinical laboratory: A longitudinal analysis of specimen identification errors. *Archives of Pathology & Laboratory Medicine, 130*, 1662–1668.

Wiener, R. S., Wiener, D. C., & Larson, R. J. (2008). Benefits and risks of tight glucose control in critically ill adults: A meta-analysis. *Journal of the American Medical Association, 300*, 933–944.

Chapter 2
Laboratory Test Utilization

Christopher McCudden

Introduction

One aspect of laboratory medicine that seldom gets presented to training physicians and patients is laboratory test utilization. Effective laboratory test utilization is a strategy for performing appropriate, cost-effective diagnostic analysis without compromising quality. It can be distilled to considering what test to order, how often to order it, and on what type of patient it should be ordered on with consideration not only of diagnosis, but also resources. Test utilization does not always equate to fewer laboratory tests. While cost is often the focus as healthcare expenditures continue to rise, test utilization is also a patient safety issue. Inappropriate tests can directly result in unnecessary blood loss and may contribute towards unnecessary procedures, follow-up testing, and may even cause harm. Test utilization is also linked to patient care directly, where selection of the appropriate lab tests aid in diagnosing or monitoring a disease more accurately. This chapter outlines the importance of appropriate laboratory utilization, identifies what kinds of tests are more expensive than others and why, and the strategies that physicians and laboratories can use to ensure effective test utilization.

C. McCudden, Ph.D., DABCC, FACB, FCACB (✉)
Clinical Biochemist, Division of Biochemistry, The Ottawa Hospital Associate Professor, Department of Pathology and Laboratory Medicine, University of Ottawa, 501 Smyth Rd., Ottawa, ON, Canada, K1H 8L6
e-mail: cmccudde@uottawa.ca

R. Molinaro et al. (eds.), *Clinical Core Laboratory Testing*,
DOI 10.1007/978-1-4899-7794-6_2

Importance of Laboratory Test Utilization

The current trajectory of healthcare costs is unsustainable. The reasons for this are myriad and include expensive technological advances, the aging population, the inefficiency of healthcare delivery, and profit-driven motives. To effectively deliver evermore expensive care to an ever-increasing population of unhealthy people it is essential to reduce waste. In the laboratory, waste occurs in the form of inappropriate tests. Waste reduction means less frequent use a given test, substitution of an expensive test with another less expensive one with similar diagnostic accuracy, or waiting for the result of one test before ordering another (reflex or sequential testing). In the context of a global hospital budget, effective test utilization could mean getting a series of laboratory tests instead of more expensive imaging tests or procedures (e.g., blood markers of heart failure, such as the natriuretic peptides, instead of an echocardiogram for assessment of heart failure). The need for effective laboratory utilization is acute and many laboratories integrate utilization into their quality management program. It is essential that laboratories, patients, and physicians all play their part in effective resource utilization to maximize care delivery and ensure a sustainable healthcare system.

Most Expensive Tests

A first step towards effective utilization is understanding which laboratory tests are the most expensive and why (Fig. 2.1). It is uncommon that patients and even many ordering physicians actually know how much laboratory tests cost. This stems from the sheer volume of tests (there are thousands) as well as the fact that laboratories don't widely inform users as to their cost. Indeed the actual costs are both complex and variable when you consider different payers and reimbursement as part of a hospital encounter. The complexity of healthcare reimbursement affects the laboratory as much as any other specialty, making the concept of shopping around for the "cheapest" test a challenge. The costs may or may not include overhead of staff and facilities and the reagents themselves. It's also not uncommon for the billed price to differ significantly from the true cost to run the test. That said, the true cost of reagents will also be wildly variable depending on the test volume and the contract negotiated between the vendor (companies that sell the equipment and tests) and the laboratory. Consider the potential difference in the price of a test for a laboratory that does 10 million tests/year with a lab that does 100,000 tests/year. So while a detailed universal table of test prices is not readily achievable, we can indicate what types of tests are most expensive than others (Table 2.1).

In general, genetic tests are the most expensive and may cost in excess of several thousand dollars for a single set of results. Genetic tests are expensive because they rely on cutting edge technology (high throughput DNA sequencers) used to do the analysis, relatively expensive reagents (enzymes, oligonucleotides), and in some cases, the cost of patents or licensing fees associated with the tests.

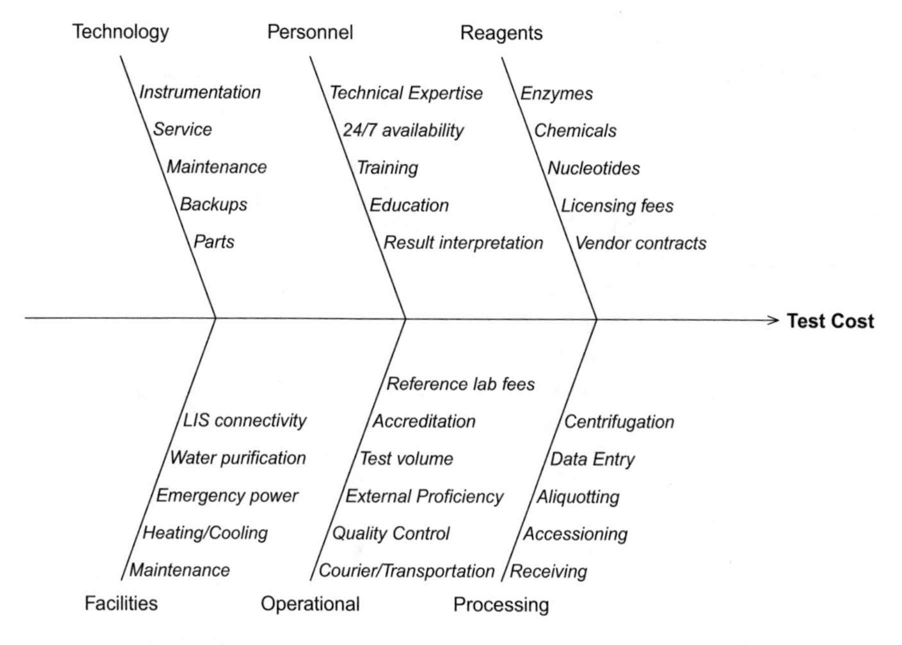

Fig. 2.1 Contributions of various factors to the cost of a given laboratory test. Billing models are often unhinged from the components that go into a test, such that utilization initiatives may have a difficult time estimating the financial impact of changing test volumes

Table 2.1 Most expensive laboratory tests

Test Category	Examples
Referral tests[a]	Chromogranin A, vitamin C
Genetic tests	CFTR, hemochromatosis
Mass spectrometry/chromatography	Tacrolimus, vitamin D
Immunoassays	Troponin, PTH
Point of care tests	Glucose, blood gases

[a]May be of any test methodology

Other expensive tests are those involving chromatography and/or mass spectrometry. Mass spectrometry and other chromatography are expensive because of the technical nature of the testing, which requires more specialized personnel with a unique skill set, and the cost of cutting edge instrumentation. A single moderate performance mass spectrometer costs between $250,000 and $750,000; a mass spectrometer can be used for quite a number for different tests, but because of operational efficiency and technical requirements they tend to be focused on a single testing area, such that the testing menu on any given mass spectrometer is actually quite small.

Immunoassays are another relatively expensive type of test, costing between $1 and $100 dollars per test depending on the type. Costs again include the price of reagents (antibodies and labeling reagents) and in some cases antibody licensing fees. These antibody-based reagents make immunoassays more expensive than routing chemistry tests, which rely on inexpensive chemicals and enzymes which are more easily generated in bulk quantities. Immunoassay is typically done on medium to large specialized instruments, although they may be done on small instrumentation on using high volume automation.

Finally, another class of tests that are generally more expensive than routine high volume chemistry tests are those done at the point of care. While point of care tests are excellent at reducing turnaround times in most settings, they are inevitably more expensive than those done on a high volume chemistry analyzer. Consider a simple glucose test, which costs a dime or two in a central laboratory, but 5–10 times more at the point of care. If speed is not of the essence, then the point of care test is not effective test utilization (POCT accuracy is another challenge entirely and is discussed in Chap. 1).

Technology aside, the most significant factor in determining the cost of a test is where it is done. Referral tests cost much more than those done on-site. While physicians may be unaware of which tests are done where, laboratories usually provide some information as to tests that are available on-site. Failing that, it can be inferred that a rare or low volume test (one not clearly named in the test menu) is going to cost more. This is a result of courier costs and the nature of the fee-for-service business that is referral testing. Many large labs will supplement their revenue by running tests for other labs at a premium. Common examples of referral tests are heavy metals, porphyrin testing, rare endocrine tests, genetic tests, and vitamin tests.

Sources of Inappropriate Tests

Appreciating the origins of inappropriate tests is the next step in mitigating waste. There are a variety of reasons that some tests are considered inappropriate ranging from technological advances and new clinical evidence to transcription errors.

One driving force in this area is patient demands and education. As patients have access to more information, they tend to want to have a say in their own care and the tests they perceive as being of use for them. While having an educated patient is a good thing, a problem arises when they have only some knowledge of a test and its use or when they get their information from dubious sources, such as online support groups or discussion forums. Physicians may be reluctant to say no to patient requests in the interest of moving them through the system, avoiding confrontation, or considering their psychological well-being. Some of these requests, such as vitamin D tests, are clinically benign, but do have an impact on healthcare costs (as described above, vitamin D is an expensive test as it is done by mass spectrometry, immunoassay, or as a referral test). In other cases, patient demands for tumor marker testing could lead to risky investigations due to their poor performance in a cancer screening setting. There are effectively no tumor markers that have a performance good enough for screening the general population (this includes PSA).

Physicians too are sometimes the source of unnecessary tests. This is particularly true for residents who may fear reprisal from an attending physician in cases when they didn't get a key test; the shotgun approach ensures they have all the results their supervisors would want. Indeed medical students are often taught from a perspective of considering everything in a differential and hypothetically testing for each. In reality there is little time for the nuances of test utilization in a packed medical school curriculum or even residency or fellowship. At the other end of the spectrum is the highly experienced physician (read old school) who may continue to order out-of-date (Jurassic) tests, which have over time been proven to offer little in the way of diagnostic information. The cardiac marker CK-MB is a prime example of a test that has been supplanted with a newer one (troponin) (Saenger & Jaffe, 2008). In other cases, generalists may find it difficult to workup rare or complex disorders and either order the wrong tests or order every test that seems to fit the bill in an effort not to miss something. Porphyrias are a good example of where it may be difficult for those unfamiliar with this rare set of hemoglobin synthesis disorders to order appropriate tests. In porphyrias, sequential/reflex testing is used in combination with clinical findings and is the best practice. In this case, physicians benefit from consulting a laboratorian or any of the diagnostic algorithms available from their own laboratory or other reference laboratories where samples may be sent.

In fairness to practitioners, the laboratory also contributes to inappropriate test utilization. Where manual processes persist, there are all too common occurrences of transcription errors leading to the wrong test at the wrong time; a classic example is daily hemoglobin tests getting translated into daily HbA1c tests (HbA1c should be ordered every 3–6 months at the most). Standing orders can also be a source of overuse. In this instance, a series of tests are repeated on a patient for an inordinate amount of time due to transcription errors rather than desired physician orders.

How Do Laboratories Control Test Utilization?

Given the necessity of test utilization, laboratories have developed several strategies for effective testing. These range from controlling the availability of orderable tests to providing physician report cards that quantify the type and volume of tests used by individual physicians. There are many considerations from the laboratory standpoint with respect to implementing test utilization strategies beginning with physician engagement, such that both parties develop a partnership to eliminate waste. Below are some strategies that are used by laboratories to improve test utilization and some tips to implementing these processes both from the lab and physician perspectives.

Pre-analytic

A simple and effective method to reducing ordering of unnecessary tests is to make them unavailable. To do this, test order forms or requisitions (electronic or paper) need to be maintained with the most up-to-date information. Obsolete tests should be removed and rare or esoteric tests should not be prominent. This simple strategy is arguably the most effective approach and with the exception of changing requisitions, is easy to implement. Physicians should be aware that rare tests are still available, but that there may be other better, quicker, and cheaper tests on hand. When a rare test is needed, the laboratory can help navigate special collection requirements or transport.

Another approach is to organize tests by disease state or appropriate ordering patterns rather than alphabetically. If the appropriate tests are groups together they are more likely to be ordered correctly. While having tests ordered as a group together makes sense, it is also useful to avoid bundling items together in a single order. It is now becoming common to separate out chloride from electrolyte orders, as it is clear that it is not needed with every assessment of potassium and sodium. Thus, a single checkbox yielding a full panel of tests is likely to drive inappropriate orders even if it's only an inexpensive test set. Practitioners need to consider: "Is every part of a basic metabolic panel necessary multiple times a day for an inpatient?" By having each test be considered independently, excess orders can be reduced. Along these same lines, it is important to review standing orders and limit their duration. We've observed instances of standing orders that outpatients faithfully follow where there is no clear indication for repeat testing. This can happen in cases where a lipid profile was ordered for a single visit, but somehow becomes a quarterly standing order which neither the physician nor patient needs. On one memorable occasion a patient had religiously had their blood collected for a panel of tests using the same requisition for more than 3 years. The requisition was a crumpled, stained, obsolete template, with a set of handwritten amendments, some of which appeared to be by the patient themself. The requisition had originated from a single presurgery encounter.

Besides monitoring what is ordered and keeping an eye on the existing menu, laboratories also carefully evaluate new tests before implementing them. This requires a conversation between requesting physicians and the laboratory leadership where they consider the clinical utility, test performance, availability of the technology, frequency and volume of orders, turnaround time requirements, and cost among other basic parameters that establish a business case for the test. Done poorly, such a review process is enough to inhibit new test implementation. Laboratories need to carefully review things they intend to insource, but also ensure they aren't a barrier to implementation of appropriate testing. Clinical investigators may be able to kick-start this process if they have some funding for a pilot project that would allow for earlier adoption and a hands-on assessment of the logistics and clinical efficacy of a newer test. In the absence of financial sense, it is unlikely that a new test will be adopted without at least some solid clinical evidence.

Another useful strategy is user education. Where testing modalities have changed or new information becomes available, laboratories can serve healthcare providers by providing educational material in the form of algorithms, practice guidelines, and publications. As medicine is a continuing educational and evolving process, such material can inform test users who will often want to understand and dissect the reasons for test utilization. In our experience, some informed users may become champions of test utilization themselves and influence peers in their clinical practice. It is also possible to find residents to be energetic advocates for appropriate testing who are willing to get involved in educational programs for their peers.

Perhaps the most draconian of strategies is direct restriction of tests. Laboratories often have a list of restricted or review tests that knowledgeable individuals review as they are requested. This may be implemented for low volume or expensive referral tests where the physician may be contacted to ensure that the request is not an error and that the result will be of clinical utility. This can be seen in cases where emergency physicians places orders for vitamin or drug tests that will not be available for weeks, potentially limiting their utility in an acute care setting. Other review processes may include complex tests that should only be ordered by area specialists, and require a particular set of clinical symptoms or criteria before they will be useful. While this may be perceived as confrontational or second-guessing, the underlying intent is to control use of resources so that they are available when they are really needed. Physicians can navigate this process by ensuring the tests they order are appropriate, and if they are recurring, can discuss their practice with the laboratory to ensure the process is not a barrier to patient care. To this end, physicians in some departments may have special dispensation for some rare or expensive tests because they are known to be part of their appropriate practice. For example, it doesn't make much sense for someone outside of neurology or oncology to order paraneoplastic testing.

Analytic

There are additional strategies that can be implemented even after a test is ordered. One effective strategy is reflex testing. This refers to sequential testing, where the results of one test inform the next. A great example of this strategy was published for ionized calcium (iCa$^{2+)}$ (Baird, Rainey, Wener, & Chandler, 2009). In that report, orders for ionized calcium on inpatients were abundant, leading to rampant dosing with calcium gluconate for an apparent epidemic of hypocalcemia in the inpatient population. An analysis of >50,000 orders showed that the treatment was not effective as the authors observed iCa^{2+} concentrations return to normal as patients recovered irrespective of the treatment. Implementation of a reflex testing algorithm where iCa^{2+} was only done in patients with abnormal total calcium concentrations curtailed both iCa^{2+} and calcium gluconate supplementation without any effect of patient outcomes. This particular example benefited not only lab utilization, but also pharmacy. Reflex testing can decrease costs and increase the diagnostic performance of tests by using Bayesian probability to advantage. This mimics a physician's thought process where they consider the likelihood of various diagnoses and proceed with a set of tests based on the probability of each. Formalization of such algorithms can ease the burden of complexity that surrounds difficult diagnoses for physicians and facilitate a linear approach to diagnostic decision-making. These algorithms often make interpretation of the results easier because of the applied logic. To be sure, algorithms can't replace critical thinking yet, but they can certainly improve the accuracy of some diagnoses and improve resource utilization.

Postanalytic

Test utilization is also influenced by the way laboratories report test results. While there are differences in format and presentation, all laboratory reports must contain certain elements as mandated by regulatory bodies (for example, the Clinical Laboratory Improvement Amendments, discussed in Chap. 1). A poorly designed laboratory report may not be easy to read or understand, leaving the clinician with more questions than answers. In such cases, physicians will be more likely to follow up with additional tests if they are unable to rule-in or rule-out a diagnosis. Thus, it is good practice to add additional information to assist the clinician in the interpretation of laboratory test results. These may include references to clinical guidelines, suggestion for follow-up steps or tests, and clear interpretative comments. These comments are often created in consultation with physicians to ensure they are informative to the intended audience. The use of visual cues, such as highlighted or variably colored results, is often underutilized in laboratory reports. This may be due to technical limitations (many laboratory information systems are old and do not allow for sophisticated reporting) or all too often, the laboratory is not involved in the discussion of how results transmitted from their system will look in the electronic medical record (EMR). Where reports or interpretative comments are not informative, you as the laboratory consumer should contact the laboratory to help improve the report. Physicians can help direct this process by making requests to both the lab and those responsible for the EMR, should the lab not be able to directly implement change. In short, presentation of laboratory results requires input from both the laboratory that produces them, and those that use the results.

Systematic Review

While all of above approaches can be effective, it is helpful to have data that can be used to identify test utilization patterns and monitor the effect of any changes made to reduce inappropriate testing. This requires comprehensive data regarding the test volumes, test orders, and result of each test. More data permits additional investigation, for example, utilization by physician specialty or by patient diagnosis. Such a systematic approach can facilitate a true measure of which of these programs are effective. It also allows for reports to be generated by test, physician, or department, which can be used to inform both administrators and physician leadership who may effect change. Reports may include focused analysis of the most expensive or most commonly misordered tests. These can also be coupled with literature, such as the latest evidence-based medicine, to educate test users in the context of contemporary clinical practice guidelines. An overview of test volumes and use may also inform operations from the standpoint of bringing in high volume tests to decrease their cost, or to change when a test is run based on order demands.

The keys to systematic reporting of test utilization are accurate data, an understanding of the processes, and the ability to present data in an informative way that accounts for the nuances of a complex system. Clinical laboratorians are adept at these tasks, and may serve as contacts for additional clinical or epidemiological questions.

How Can Physicians Influence Test Utilization

As evident from the above sections, the laboratory has a significant influence of test utilization. However, physicians themselves can in turn influence the laboratory through communicating their perspective and clinical expertise to the laboratory leadership.

Physicians can also affect test utilization through education. They can avail themselves of diagnostic test uses and clinical practice guidelines. They may consider asking laboratorians to share their expertise on committees that design clinical pathways or diagnostic algorithms.

Uniquely, physicians are also able to educate patients. They may inform patients of the risk of tests that aren't indicated or share primary literature with those who are willing and able to consume it. Were patients to have better information sources, they would be better informed and be more likely to request or accept the appropriate diagnostic test. Finally, physicians can influence their peers. There is strong literature that supports how physician practice is affected by others, where social science can inform change management (Wong et al., 2013).

Summary

Test utilization is an important aspect of laboratory medicine that is becoming an integral part of overall quality. While laboratory tests provide the biggest "bang for the buck" in the healthcare system, there are instances of overuse and opportunities to conserve resources. An informed physician can not only use tests effectively on their own, but also be part of the overall initiative where they help change practice. Any test user is likely to benefit from the information available by the lab. The concerted use of appropriate tests can not only increase the accuracy of blood test diagnostics, but also allow for the availability of resources needed for rare and complex cases. Test utilization is not just saving money, but improving patient care.

References

Baird, G. S., Rainey, P. M., Wener, M., & Chandler, W. (2009). Reducing routine ionized calcium measurement. *Clinical Chemistry, 55*, 533–540.

Saenger, A. K., & Jaffe, A. S. (2008). Requiem for a heavyweight the demise of creatine kinase-MB. *Circulation, 118*, 2200–2206.

Wong, B. M., Kuper, A., Hollenberg, E., Etchells, E. E., Levinson, W., & Shojania, K. G. (2013). Sustaining quality improvement and patient safety training in graduate medical education: Lessons from social theory. *Academic Medicine, 88*, 1149–1156.

Chapter 3
Before the Lab Tests Run: Preanalytical Issues in the Clinical Laboratory

Deanna Franke and Marjorie BonHomme

"Garbage in, garbage out" is a familiar phrase from the field of computer science. The concept surrounding the phrase is simple. Despite the most complex information processing systems, computers will unquestioningly process the most ridiculous of input data, "garbage in," and produce ridiculous output, "garbage out." Commonly used in other fields, including laboratory medicine, this phrase serves as an important reminder that in the context of healthcare, inaccurate data input—"garbage in"—leads to misleading results—"garbage out". So before the laboratory can perform testing it is imperative that inherent, preanalytical issues are understood and identified. This chapter focuses on specimen integrity in the preanalytical phase of testing and discusses how the laboratory strives to link the quality of incoming specimens to quality results.

Physicians rely on the clinical laboratory to provide accurate, patient-specific data for general well-being and diagnostic and prognostic assessment of disease. In order to generate high-quality patient data, the clinical laboratory requires high-quality patient specimens. When specimens are compromised, "garbage in," patient care is compromised, "garbage out." The overall quality of patient specimens and their appropriateness for laboratory testing are determined during the pre-analytic phase of laboratory testing. The entire healthcare team including physicians and consulting medical experts, nurses, pharmacists, and laboratorians are responsible for this phase of testing.

D. Franke (✉)
Lab Operations, Carolinas HealthCare System, PO Box 32861, Charlotte, NC 28236, USA
e-mail: Deanna.Franke@carolinashealthcare.org

M. BonHomme (✉)
Lab Operations, Acutis Diagnostics, 728 Larkfield Rd, East Northport, NY 11731, USA
e-mail: Marjorie@acutisdiagnostics.com; mbonhomme231@live.com

R. Molinaro et al. (eds.), *Clinical Core Laboratory Testing*,
DOI 10.1007/978-1-4899-7794-6_3

Specimen Quality Is the Key to Effective Diagnosis

Human beings, in our infinite diversity, have remarkably similar biochemical profiles when we are healthy. Deviations from our "healthy" biochemical profiles require further investigation in symptomatic patients. Specimens are snapshots of a patient's condition. If the specimen is not properly handled, then the diagnosis may be delayed, mismanaged, or missed entirely.

Take for example a case of a 5-year-old child who visits his pediatrician's office for what appears to be a cold with chest congestion, persistent fatigue and nausea, and a few episodes of vomiting. After meeting with the patient and his mother, the pediatrician orders a basic metabolic profile and a few other tests. The in-office phlebotomist performs a venipuncture and collects a serum sample at 8:30 AM. The pediatrician is a member of a bustling medical practice that includes pediatricians, physician's assistants, registered nurses, licensed practical nurses, in-office phlebotomists, receptionists, and a practice manager. The practice manager recently contracted with a new laboratory service provider. However, this information and subsequent process changes were not communicated. No one takes notice when the patient's requisition and sample are placed in the courier bin belonging to the previous laboratory service provider. Due to this oversight, the specimen is not sent to the laboratory until 4:30 PM. The pediatrician reviews the laboratory results the next day and finds nothing remarkable. Two weeks later, the patient is taken to the emergency room after being found on the floor and unresponsive. It was determined that the patient was experiencing diabetic ketoacidosis, a serious condition that can lead to death if not diagnosed and managed.

The warning signs of diabetic ketoacidosis and a diagnosis of new-onset diabetes likely were missed 2 weeks prior. Why? It turns out that glucose decreases in blood at the rate of 6–7 mg/dL/h (Chan, Swaminathan, & Cockram, 1989). The rate of glucose decrease is even faster in specimens taken from patients with infections due to increased microbial activity or leukemia due to increased number of metabolically active white blood cells. By the time the specimen was received, processed, and analyzed in the laboratory, the specimen's glucose concentration dropped considerably, on the order of 60–100 mg/dL. The integrity of the specimen was compromised which allowed its glucose concentration to go from abnormally high to normal. The laboratory did not note the significant delay in processing and testing from time of collection and the pediatrician was unaware of the delay when the laboratory results were reviewed. Overall, these delays put the patient at risk of experiencing complications like cerebral edema—the most common cause of diabetes-related morbidity and death (Bohn & Daneman, 2002; Edge, Hawkins, Winter, & Dunger, 2001; Glaser et al., 2001; Wolfsdorf, Glaser, Sperling, & American Diabetes Association, 2006). This case is meant to emphasize three things. First, patients and physicians depend on quality laboratory results for early diagnosis, as well as prevention of disease complications which has great relevance in limiting hospitalizations. Second, quality laboratory data begins with quality specimens.

Lastly, quality specimens and communication of any irregularities throughout the total test process, from order to result, are the responsibility of the entire healthcare team.

What Is the Total Test Process?

First coined by Lundberg, the "total test process" is best described as a cycle (Lundberg, 1981, 1999). The brain-to-brain loop begins when the clinician first thinks to order and then places an order for a laboratory test. The patient specimen is then collected and transported to the laboratory, where the specimen is then processed and analyzed and results are communicated back to the clinician. The clinician then interprets and takes action based on the reported laboratory results. Altogether, this brain-to-brain cycle from order to action represents how clinicians commonly define turnaround time and can be further segregated into three phases: preanalytical, analytical, and post-analytical. Discussion of analytical and post-analytical phases can be found in Chaps. 5 and 6.

What Does Preanalytical Mean?

The preanalytical phase of testing encompasses all the steps beginning when the physician thinks about ordering a test until the specimen is ready for analysis. Most of the activities that happen within the preanalytical phase are near patient and outside of the laboratory. Unfortunately, nearly 60 % of errors are made during the preanalytical phase of testing (Ernst, 2005). Most of these errors can be attributed to the wrong test being ordered, improper collection technique, and inappropriate specimen handling by the entire healthcare team.

According to the Clinical Laboratory Improvement Amendment of 1988 (CLIA '88) the laboratory is responsible for specimen integrity during all phases of testing. Historically blood collection has been centralized to phlebotomists under the direction of the laboratory. However, as described in Chap. 1, under mounting financial pressures healthcare has been forced to restructure itself, commonly resulting in the decentralization of blood collection from laboratory-based phlebotomists to nurses, physicians, and other allied health professionals. This decentralization of blood collection out of the laboratory has forced many healthcare professionals to become proficient in specimen collection tasks with minimal formal training. Fortunately, in accordance with CLIA '88 regulations, clinical laboratories commonly support education and training efforts as well as publish a specimen collection manual that is readily available in areas where patient specimens are collected. These manuals provide detailed instructions and information to minimize errors and ensure quality results. Healthcare professionals involved in test ordering and specimen collection should familiarize themselves with their laboratory's collection manual and consult their laboratory when in doubt.

Preanalytical Issues During Specimen Collection

Clinical laboratories receive a multitude of specimens daily. To manage this, large-volume laboratories commonly turn to automated systems to access, process, and deliver specimens to analytical testing lines. These high-throughput, automated systems work well when they are encoded with logic-based, clear rule sets and instructions. Likewise, lower volume laboratory spaces, which may be devoid of automated systems, generally rely on hardwired, standardized manual processes to minimize human intervention as much as possible. In either case, deviations from hospital- and laboratory-approved specimen collection protocols may compromise specimens and limit their use in analytical test systems. In these cases, the laboratory must intervene to evaluate if the specimen is acceptable for testing or reject the specimen. Both situations lead to increased healthcare cost, depletion of resources, delayed turnaround times, and additional discomfort to the patient if a second specimen collection (venipuncture, spinal tap, etc.) has to be performed. The following subsections will systematically present a review of the preanalytical phase of testing variables and discuss how these variables, if left uncontrolled, could impact the quality of laboratory results.

Ordering a Laboratory Test

Key steps to ordering the desired laboratory test are to ensure that both the matrix (e.g.: whole blood, plasma, serum, cerebral spinal fluid (CSF), urine) and analyte or test (e.g.: glucose, potassium) are defined. A test order for total protein is not a complete description of a test if the laboratory offers total protein testing in both serum and CSF. Keep in mind that just because your laboratory offers serum total protein testing does not automatically mean that they can measure total protein in CSF or any other fluid. Before placing an order, a healthcare provider should consult their laboratory's test directory and specimen collection manual. These resources are usually available electronically within the information system used at the point of order or on the hospital network and clearly define the laboratory's test menu and state the accepted specimen matrices. Consulting these resources and placing the test order correctly ensure that specimens are properly handled and tested, and results are reported with the correct units and reference range.

In general, the breadth of tests performed in the laboratory (in-house) is a direct reflection of the size of the healthcare institution as well as the supported clinical service lines (e.g.: pediatrics, transplant, oncology, orthopedics). Clinical laboratories routinely affiliate with other laboratories and send out the more non-routine and specialized, esoteric tests. If a test is not found in the laboratory test directory, then a healthcare provider should call the laboratory to find out if the desired test can be performed at an affiliated referral laboratory. It may also be important for the provider to obtain specific ordering information from the referral laboratory; however, the institution's clinical laboratory makes the final decision as to what tests can be sent out and to which laboratories. There are a number of factors that are considered when selecting the appropriate referral laboratory for desired testing, including but not limited to the clinical utility (what is the predictive value of the test), turnaround time (will the result be available by the time the medical team needs to act), and cost (will the laboratory be able to bill and get reimbursed). Usually if the clinical need is apparent, cost becomes less of a factor as hospitals generally put the patient first and will absorb the cost. For novel, rarely ordered, or expensive tests, providers should contact the laboratory's medical director or medical/scientific specialist (e.g.: clinical chemist, clinical microbiologist, hematologist). These professionals will be able to help identify the appropriate referral laboratories as well as expedite specimen collection, processing, and transport. Collecting a specimen in the absence of laboratory consultation may result in the specimen being rejected and delaying patient care. Therefore, to prevent unnecessary blood loss and discomfort for the patient and ensure quality patient results, providers should actively engage the laboratory during the test ordering process.

Identifying the Patient

The most important step to specimen collection is correct patient identification. Failure to correctly identify a patient can result in treatment errors, leading to serious injury or even death. To avoid these errors, healthcare professionals responsible for specimen collection should first read, understand, and follow institutional protocols and policies. Once appropriate orientation and training are completed it is time to interface with the patient.

Appropriate patient identification requires the use of at least two identifiers—this ensures that the right specimens are collected from the right patient. The first identifier is typically the patient's name—first and last. For patients within the hospital (in-patients), the second identifier is usually the assigned medical record number, the number used by the hospital to document medical history, and care during the hospital stay. Healthcare professionals responsible for specimen collection should never rely on signage on the door or above the bed with the patient's name since patients can be moved between rooms, wards, or floors. Armbands are commonly used in healthcare institutions to identify patients, contain both the patient name and medical record number, and should always be used as the source of truth for proper patient identification. Specimens should never be collected from patients without armbands. In these cases, the armband must be reapplied before specimen collection. For patients outside the hospital (outpatients), healthcare professionals commonly use the patient date of birth, social security number, or address as a second identifier. No matter where specimens are collected, any discrepancies regarding patient identity should be reconciled prior to specimen collection. Once the patient has been properly identified, healthcare professionals should also reconcile test orders by reviewing written requisitions or printed specimen labels. Post-collection activity should always include labeling the specimens prior to leaving the patient's side with the appropriate labels and collector's initials, date, and time. Refer to the subsection on labeling specimens for more details on patient-centered care during the specimen labeling process.

Creating a Safe Environment on Initial Encounter

At initial encounter, the healthcare professional responsible for specimen collection has a great opportunity to create a safe and welcoming environment for the patient. The collector should greet the patient with a smile, make eye contact, and state the purpose of the visit. After introducing himself or herself it is important to be attentive to the patient and ask the patient to state their name. Since patient name is one of the primary identifiers used to ensure appropriate collection, patients should never be asked: "Are you Mr. Smith?" Some patients, for a variety of reasons, will answer yes to any question posed to them by a healthcare professional. This is important to remember because patients may not hear well or are too nervous to pay attention, some have language barriers, and others believe that out of respect, healthcare professionals should never be questioned.

Healthcare professionals should also be prepared to listen to patient concerns and blood collection preferences, and accommodate any reasonable request. Venipuncture can be quite stressful for some patients. Stress can raise the blood concentrations of adrenal hormones, fatty acids, lactic acid, and white blood cells. A patient can be so distressed that there is pressure to hasten the venipuncture procedure, resulting in collection of less than the optimal blood volume required and potentially causing hemolysis of the blood specimen. Altogether, these preanalytical variables can significantly impact the ability of the laboratory to provide accurate, diagnostic information to the physician. If the patient appears to be stressed, talking to the patient throughout the process or asking for assistance from another healthcare provider or family member to hold the patient's free hand is ideal. Institutional policies should be followed when obtaining specimens from patients who have language or communication barriers or who are unconscious. Unconscious patients may be aware of the presence of others and as such should be greeted and communicated to just as if the patient were conscious.

Preparing the Patient

In advance of specimen collection and when appropriate, patients should receive pretest instructions aimed at reducing the impact of preanalytical variables on laboratory tests. These instructions may include modification of diet (e.g.: fasting or eliminating certain foodstuffs), abstinence from medications (prescribed and over-the-counter) and supplements (herbal and nutritional), and avoidance of strenuous exercise and stress. Deviations from these pretest preparation instructions should be verified and noted by the healthcare professional at the time of collection as noncompliance may lead to inaccurate results.

Modification of Diet: Due to religious and personal beliefs, individuals may have different definitions of fasting. A healthcare provider should take the time to define fasting for their patients. Fasting is usually defined as refraining from food and liquids, except water, for at least 10–12 h, but not more than 16 h. Some populations however, like the very old and very young, may not be able to comply with this request. In the absence of an appropriate fasting period, two common test results that are increased include glucose and triglycerides. These analytes can remain elevated for 4–8 h following a meal due to intake of food high in fat, carbohydrates, and simple sugars. Other test results that will increase following a meal include insulin and liver enzymes, while potassium, ionized calcium, chloride, and phosphate may decrease (Young, 2012). While the effect of fasting varies depending on body mass, it is well recognized that prolonged fasting, due to mobilization of lipids from adipose and muscle tissues to protect the body from the effects of starvation, will also cause an increase in triglycerides and fatty acid concentrations. Fasting beyond 48 h can result in a significant increase in bilirubin as well as decreases in the C3 component of complement, prealbumin and albumin (Narayanan, 2000; Statland & Winkel, 1977).

Fasting noncompliance is a significant preanalytical issue. An indication that a patient has not fasted or regularly consumed foods high in carbohydrates and saturated fat is a specimen that appears lipemic (cloudy). High levels of lipemia will interfere with tests that utilize spectrophotometric techniques for detection and quantitation. In addition, with greater recognition and concern for the development of prediabetes and diabetes, fasting glucose and oral glucose challenges are key diagnostic indicators that healthcare providers use to screen and assess their patients (American Diabetes Association, 2016). Important pretest instructions for oral glucose challenges include eating a well-balanced diet that includes 150 g per day of carbohydrates for 3 days and reporting to the test and draw site in a fasting state. Patients undergoing glucose load challenge testing should also avoid nicotine, caffeine, and chewing gum because they stimulate digestion and alcohol because it inhibits glucose metabolism. If the patient is not appropriately fasting or does not follow pretest instructions, patients may be misclassified or asked to return for other screening tests unnecessarily. Refer to *Timing of Collections* for more discussion.

Medications and Supplements: When taking medications or supplements known to interfere with laboratory tests, physicians commonly request patients to discontinue use for a specified length of time (e.g.: 24–72 h) prior to specimen collection. The length of time required is dependent on prescribed dose, clearance, as well as the test and specimen of choice (blood vs. urine). Depending on the patient condition and medication requirements, a physician may choose to place the patient on another drug. It is important to note that patients should never discontinue medications unless instructed to do so by their physician. Failure to follow pretest instructions in regard to stopping medications or abstaining from supplements can be a major source of preanalytical variability in laboratory testing.

Consider the following case (Scenario adapted from Connor, Hermreck, & Thomas, 1988). A 51-year-old woman with a 10-year history of anxiety, recent antianxiety medication change, and new-onset hypertension returned to her physician with concerns that she may have a pheochromocytoma, an adrenal hormone-secreting tumor that may precipitate life-threatening hypertension. The recent laboratory work ordered by her physician demonstrated a moderate elevation in plasma normetanephrine concentration, nearly three times the upper reference limit. The patient reminded her physician that she recently lost her twin sister due to this condition and risk analysis through genetic testing confirmed that she too carried the genetic mutation for succinate dehydrogenase subunit B, placing her in a higher risk category for developing a pheochromocytoma or paraganglioma. Since elevated metanephrines (metanephrine and normetanephrine) are reported to have a very high positive predictive value and specificity for these conditions, it appears to be an open-and-shut case. However, upon rereview of her medications the physician remembered that he requested this patient to discontinue use of her newly prescribed antianxiety medication, venlafaxine. Venlafaxine, a selective serotonin and norepinephrine reuptake inhibitor (SSNRI), works to increase normetanephrine concentrations by decreasing the reuptake of norepinephrine and as such should be avoided for at least 48–72 h prior to blood collection for normetanephrine determinations. When asked, admittedly the patient confessed that she forgot to stop taking venlafaxine. The physician reordered the testing, lowered the venlafaxine dose, and reminded her to follow the provided pretest instructions to withhold for 72 h prior to getting her blood drawn. The patient returned to her physician 3 weeks later, and upon review, the patient's plasma normetanephrine concentrations and blood pressure had normalized. This patient did not have a pheochromocytoma but was experiencing high blood pressure as a side effect of her medication that was corrected with the dose adjustment. The initial laboratory work demonstrating moderate elevations in normetanephrine was due to the patient not abstaining from taking her prescribed venlafaxine as instructed.

There are a number of other medications that affect metabolism, clearance, and movement of catecholamine family analytes that should be discontinued prior to blood collection for certain laboratory tests. Like SSNRIs, monoamine oxidase inhibitors (MOIs) increase normetanephrine concentrations but through a different mechanism of action. MOIs prevent the metabolism of norepinephrine. Sympathomimetic agents and adrenergic-receptor blockers should also be avoided. Sympathomimetic agents, like caffeine, nicotine, ephedrine, and pseudoephedrine, increase catecholamine concentrations due to their ability to stimulate adrenergic receptors. Adrenergic-receptor blockers significantly decrease catecholamine concentrations (Connor et al., 1988).

An example of a supplement demonstrated to interfere with immunoassays is biotin, a water-soluble B vitamin also known as vitamin B7. Biotin is a coenzyme for enzymes involved in gluconeogenesis and the synthesis of fatty acids and some amino acids. High-dose biotin supplementation has been reported to improve symptoms of encephalopathy and peripheral neuropathy in patients with renal failure and diabetes (Head, 2006; Yatzidis et al., 1984). Biotin has also been touted as a beauty supplement to aid in the growth of strong nails and hair. Regardless of the reason, there are many people taking large doses of biotin. It has been demonstrated that the biotinylated molecules present in specimens collected from these patients may interfere with the biotin-streptavidin mechanisms widely employed in immunoassay method designs (Elston, Sehgal, Du Toit, Yarndley, & Conaglen, 2016; Kummer, Hermsen, & Distelmaier, 2016). Therefore it is recommended that patients discontinue biotin supplementation at least 8 h before having their blood drawn for an immunoassay-based test. These examples illustrate the fact that prescribed medications and supplements can be a significant source of preanalytical variability and their presence should be considered when reviewing laboratory results to prevent misinterpretation, misdiagnosis, and costly medical misadventures.

Exercise: Exercise is an important part of maintaining a healthy lifestyle; however, engaging in strenuous exercise just before blood collection should be avoided. High levels of activity and strenuous exercise can significantly alter concentrations of measured analytes in blood. Exercise depletes muscle cells of adenosine triphosphate which is required to maintain cell membrane integrity. As a result, creatine kinase, aspartate aminotransferase (AST), and lactate dehydrogenase (LD) are released from muscle cells, thus increasing their serum concentrations (Young, 2012). Exercise may also cause in vivo hemolysis resulting in the release of free hemoglobin, AST, and LD. AST and LD increase due to release from red blood cells and free hemoglobin binds to haptoglobin resulting in decreased haptoglobin concentrations (Yusof et al., 2007). Exercise has also been shown to increase the concentration of gluconeogenetic hormones, like cortisol and epinephrine, resulting in increased glucose concentrations (Young, 2012).

As illustrated by these examples, numerous types of interferences are present that can be controlled and minimized by discussion with the patient and appropriate preparation. The laboratory can serve as an essential resource to help address questions.

Timing of Collections

In addition to a patient's general condition, the timing of collection plays an important role in the ability to assess some conditions. Timed collections are commonly used to determine a subject's ability to metabolize molecules, clear medications (therapeutic drug monitoring), measure analytes that exhibit diurnal variation, and assess biochemical responses following stimulation or load challenges.

Therapeutic Drug Monitoring: Aminoglycosides are a class of antibacterial therapeutic agents used to treat infections. They are not metabolized in vivo, but are freely filtered by the glomerulus, taken up by and concentrated in the proximal tubular cells, and then cleared by the kidneys. Aminoglycosides can be nephrotoxic leading to renal impairment (Meyer, 1986) as well as ototoxic (Zheng, Schachern, Sone, & Papapella, 2001) resulting in irreversible hearing loss. Given their narrow therapeutic index and high potential for toxicity, therapeutic drug monitoring of aminoglycosides is necessary.

Therapeutic drug monitoring for aminoglycosides is commonly achieved by performing peak, random, and trough-level assessments. Peak levels should be collected between 15 min and 1 h after the end of infusion. The later the sample is collected, the more likely the result will significantly underestimate the true peak. As named, random levels can be assessed anytime during therapy. Trough levels should be collected up to 30 min prior to the next dose.

There are times when mistakes are made that can significantly change the patient's course of treatment. The patient's drug dosage can be increased, decreased, or stopped based on the laboratory result. Inappropriate collection of therapeutic drug monitoring samples during infusion or immediately after administration will overestimate its concentration. In addition, the laboratory commonly receives specimens labeled as a trough when in fact the drug infusion was started prior to the collection of the specimen. This may result in confusion, and needed therapy to be withheld, posing a significant threat to the recovery of patient. Regardless of the test order (peak, random, and trough) to prevent inappropriate changes in dosing, it is very important that the healthcare team, nurse and the blood collector, note the exact time the last infusion started and ended and the time of collection on the specimen tube.

Load Challenge: The oral glucose tolerance test is used to test for diabetes (American Diabetes Association, 2016, Jackson et al., 2016), insulin resistance (Stumvoll et al., 2000), and less often reactive hypoglycemia (Stuart, Field, Raju, & Ramachandran, 2013), and acromegaly (Carmichael, Bonert, Mirocha, & Melmed, 2009). To test a patient's ability to metabolize glucose, a solution is administered or given to the patient to consume. The solution is a standard glucose load that may be fixed or proportional to the patient's body weight. Blood is then collected at designated times, usually fasting (time 0) and 1, 2, and 3 h post-dose. Normal patients, in response to the standard glucose load, will demonstrate a spike in blood glucose levels and then quickly return to normal. However, patients with increased insulin resistance demonstrate a delayed response or slower rate of return to normal glucose levels. Adherence to the specified times and noting the time of collection on the tube or requisition form are important as these timed results are compared to expected cutoffs for appropriate evaluation and diagnosis.

Positioning the Patient

Posture is a controllable preanalytical variable in most patients. When in a standing position, gravity causes an increase in hydrostatic pressure which results in plasma water and small molecules to leak from the intravascular compartment while leaving behind cells and larger protein species. In this position, larger analytes will show a 5–15 % increase, which can be clinically significant. When lying in a supine position, plasma water returns to the intravascular compartment shifting concentrations of analytes back to the same extent.

Free fractions of metabolites, drugs and hormones, small molecules, and metal ions are not generally subject to the effects of posture; however, as alluded to, larger protein-bound fractions of these analytes can be significantly affected by body position. For example, total calcium (45 % of calcium is bound to albumin), total bilirubin and delta bilirubin (8–90 % of total bilirubin is albumin bound in patients with hepatocellular and cholestatic jaundice), and total cholesterol are all affected by the effects of posture (Young, 2012). In general, the effects of posture appear to be intensified in patients with edema due to cardiovascular insufficiency and cirrhosis.

In the outpatient setting, patients should be in sitting position for at least 15 min prior to blood collection. This may not be possible for children and needle-phobic patients. The posture of children during venipuncture will be determined by the caregiver holding them and whatever seems to reduce the child's anxiety. For patients that appear to be considerably anxious, the supine position may be preferable. This will also prevent the patient from injuring themselves should they experience an episode of syncope. Regardless of what position blood is collected, deviations from your institution's protocol should be noted and it may be necessary to document significant information on the requisition form or within the patient's visit information at collection.

Timing of Tourniquet Application

During blood collection tourniquets are used to temporarily accumulate blood to make it easier to locate the veins. Despite the reduced blood flow, enough pressure remains to force plasma water and small molecules out of the intravascular compartment into the local tissue, thereby artificially increasing some analytes. Due to blood stasis the restricted movement of larger lipoproteins like cholesterol may increase by 5 % when the tourniquet is applied for 1–2 min and up to 15 % when applied for more than 15 min (Cooper, Myers, Smith, & Sampson, 1988; Guder, Narayanan, Hermann Wisser, & Zawta, 2007). Prolonged tourniquet application also leads to local decreases in oxygen tension, which turns on anaerobic glycolysis. Under these anaerobic conditions, lactate concentrations increase causing a decrease in blood pH. Low blood pH or high hydrogen ion concentration is alleviated by moving hydrogen ion into cells in exchange for potassium. Thus prolonged tourniquet application can lead to increased potassium concentration. This potassium increase may be exaggerated further if the patient repeatedly clenches their fist as repeated muscle contractions can result in an increase of potassium up to 2.7 mmol/L (Baer, Ernst, Willeford, & Gambino, 2006; Don, Sebastian, Cheitlin, Christiansen, & Schambelan, 1990).

To decrease preanalytical issues related to tourniquet application, the tourniquet should be maintained for less than 1 min and released as soon as blood begins to flow into the blood collection device. Additionally, once a vein has been easily located, the patient should be asked not to clench their fist.

Intravenous Access Devices

Intravenous access devices deliver fluids, blood products, nutritional support, and medications directly into a patient's circulatory system. While blood collection by venipuncture will remain the gold standard, blood drawn from intravenous access devices will continue to be utilized because they provide convenient access to the circulatory system, increase patient comfort, and expedite care. While blood can be collected from central venous catheters care should be taken to maintain optimal specimen integrity.

If a patient is receiving an infusion, it is generally recommended that blood is sampled from the opposite arm or that a certain amount of time be allowed to pass after the infusion is completed and prior to collection. If blood is drawn during an infusion and from the same arm, then the sample collected will have falsely elevated concentrations of infusion-specific analytes and falsely decreased concentrations of others. Take for example the following case. A patient is receiving an infusion of 5 % dextrose monohydrate (5 g/100 mL of water or 4545 mg/dL of D-isomer glucose). A physician ordered a glucose measurement. A sample is collected and sent to the laboratory. The sample's glucose concentration is resulted as >800 mg/dL. An hour after sample collection, the laboratory calls the medical team with the critical glucose result. The medical team is shocked. The patient is not a diabetic and does not show any signs of experiencing hyperosmolar, hyperglycemia. A subsequent glucose determination, by point of care, measures 130 mg/dL. The medical team calls back the laboratory to inform them that there must have been a laboratory error. An investigation is launched to determine the cause of the discrepant results. It is determined that institutional procedures were not followed with respect to blood collections and infusions and the specimen was drawn from a vein above the infusion line and while the infusion was taking place. Remember that many "laboratory errors" occur before the laboratory ever receives the sample.

If a basic metabolic profile (BMP) was ordered instead of glucose, this scenario may have played out a little different. Let's explore this version of the same case. The results of the BMP for the compromised sample are listed in Table 3.1. The concentrations of sodium, chloride, CO_2, and calcium fall below the reference range, while those of potassium and creatinine are on the low side of normal. Assuming that a

Table 3.1 Basic metabolic panel results

Analyte	Compromised sample	True patient results	Reference range	Units
Sodium	119	142	137–147	mmol/L
Potassium	3.7	4.4	3.4–5.3	
Chloride	87	104	99–108	
CO_2	21	26	22–29	
BUN	12	15	8–21	mg/dL
Creatinine	0.8	1.0	0.7–1.1	
Glucose	>800	130	60–200	
Calcium	8.1	9.7	8.7–10.7	

prior BMP was available, sodium, potassium, chloride, creatinine, and calcium would have been flagged for failed delta checks. With this added information, the laboratory would have called the medical team to inform them that the test was canceled due to suspected contamination and asked them to recollect the sample.

For other types of nutrient infusions, waiting for a period of time is recommended. For patients receiving Intralipid® (fat emulsion) commonly found in transperenteral nutrition (TPN) preparations, it is recommended that at least 8 h elapse from the end of infusion to the collection of blood samples. For patients receiving protein solutions, like AlbuRx® or immunoglobulin G, and electrolyte solutions, a 1-h wait time is recommended. See ***Timing of Collections*** section for other discussion.

Blood drawn from intravenous access devices is convenient for healthcare professionals and patients, but collection of blood thorugh these devices has the potential to produce false laboratory results. Always follow your institution's policy for line draws in terms of flushing the line with saline, blood wastage, and waiting for a certain period of time following infusions prior to blood collection.

Selecting the Appropriate Tube Types

In order for the clinical laboratory to deliver accurate and reliable information for diagnostic use, collection of appropriate blood samples from the patient is critical GP41–A6. During the venipuncture, blood is drawn into evacuated collection tubes which fill automatically due to the premeasured vacuum in each tube (Bush & Cohen, 2003; GP44-A4 2010). There are a number of evacuated collection tubes that are differentiated via colored stoppers or plastic caps at the top of the tubes. Functionally, the color differences are a visual representation of different additives in each tube that prepare and preserve the specimen for appropriate analysis in the laboratory (GP34-A) (Table 3.2).

Collecting the Correct Sample Volume

The volume of blood collected is critical for some laboratory tests. This is especially true for samples collected in tube types containing liquid additives (see Table 3.2). Under or over filling collection tubes can significantly interfere with analytical studies due to the increased or decreased concentration of the additive. SPS aids in microbiological studies by inhibiting the natural defenses of blood, like phagocytosis, complement and lysozyme activity, and inactivating antibacterial therapeutic agents used to treat patients. Overfilling tubes may not provide the adequate concentration of SPS to inactivate cellular activity or the therapeutic agents, thereby making it potentially difficult to identify the causative microorganism. Likewise, coagulation studies require the correct ratio of the anticoagulant, citrate, to blood (1:9) for accurate test results. Collecting less than the optimal volume of blood will

Table 3.2 Common tubes used for blood collection

Cap color	Additive(s)	Purpose	Laboratory use
Yellow	SPS[a]	Whole blood/plasma	Microbiology—blood culture
	ACD[b]	Whole blood/plasma	Blood Bank, HLA phenotyping, DNA testing
Light blue	Citrate	Whole blood/plasma	Coagulation
Red	None	Serum	Chemistry, serology, donor screening
Gold	None	Serum	Chemistry, serology, donor screening
Green	Heparin	Whole blood/plasma	Chemistry
Lavender	EDTA[c]	Whole blood/plasma	Hematology
Pink	EDTA[c]	Whole blood/plasma	Hematology, donor screening
Royal blue	Heparin	Whole blood/plasma	Toxicology, trace element testing
	EDTA[c]	Whole blood/plasma	
	None	Serum	
Gray	NaF/KOx	Whole blood/plasma	Glucose determinations
	NaF/EDTA	Whole blood/plasma	
	NaF	Serum	

[a]Sodium polyanethol sulfonate
[b]Acid citrate dextrose
[c]Ethylenediaminetetraacetic acid

prolong the clotting time. This effect is more pronounced for aPTT than for PT. Reducing the ratio to 1:7 can increase the aPTT time by 2.4 s, which is a significant change in a patient's aPTT result. Over or under filling tubes used for coagulation studies could have profound effects on therapeutic dosing of anti-thrombolytic agents. Lastly, underfilling lavender top tubes, for complete blood count analysis, affects the osmotic environment leading to cell shrinkage and distorted cell morphology.

Collecting the right amount of blood for the right testing the first time is critical for effective patient care and, at times, for the protection of the community at large. Consider the following case. After getting off work from a local restaurant a 23-year-old female goes to visit a local physician's office with complaints of abdominal pain, nausea, vomiting, and jaundice. When asked about recent travel outside the US, she said that she was just in Haiti 3 weeks ago on a missionary trip with her church group. She declined pretravel vaccinations. The physician ordered a battery of laboratory tests, including a hepatitis panel (hepatitis A IgM antibodies, hepatitis B surface antigen, hepatitis B IgM core antibody, hepatitis C antibodies). Unfortunately, the patient was needle-phobic and proved to be a challenge. The healthcare professional collecting the specimens decided to hasten the process and obtained less than the requested volume of blood, hoping that the laboratory would make due as they had in the past. The laboratory performed most of the tests but the report had a comment for hepatitis A IgM that read "quantity not sufficient." The physician's office attempted to call the patient to request that she return to provide another specimen and to discuss the initial laboratory tests. However, due to the patient's now busy work schedule between two local restaurants, the office had a difficult time locating her. By the time the patient was located, she was feeling better and decided that she doesn't need further medical attention. Two months later the State Health Department contacted the physician office to inform them that the patient has been named the index patient for a hepatitis A outbreak involving over 250 individuals.

This case illustrates the importance of getting it right the first time—right patient, right specimen, and right test and that with some cases there is only one opportunity to make an impact. The outcome might have been different had the physician's office, the laboratory, and the public health department known of her acute hepatitis status earlier. The healthcare professional collecting the specimens should have sought help or told the physician of the issue. The physician should have been given the opportunity to come up with an alternate diagnostic or testing plan to prioritize. It is important to recognize that insufficient sample volumes are common amongst elderly, oncology, and pediatric patients. Careful consideration is always required to ensure that appropriate specimens are obtained so that patients get the appropriate diagnostic workup.

Labeling Specimens

High-throughput, automated laboratories utilize multiple barcode scanners to identify, access, and properly route specimens to different laboratory sections and to various instruments with minimal human intervention. Laboratory scanners are optimized to read barcodes in one orientation and within a certain distance. The ability of scanning systems to perform efficiently depends on the quality of the labels used and printer. Healthcare professionals should learn how to properly affix specimen labels to specimen tubes and to notify the correct individuals when the quality of barcodes diminishes. Poor-quality labels or improperly placed labels on specimens lead to increased turnaround times and consume resources in terms of laboratory time to reprint and affix new barcodes. Appropriately labeling specimens is an important pre-analytical step to getting quality results. With appropriate issue management and education, labeling errors can be minimized.

Handling and Transporting Specimens

Institutional policy and procedures should always be followed to minimize preanalytical specimen handling variables that may lead to inaccurate laboratory results. Within institutions, most specimens can be transported at ambient temperature and delivered to the laboratory via pneumatic tube systems or by hand directly to the laboratory. Some analytes require specialized collection and transport conditions for optimum analyte measurement. Angiotensin-converting enzyme, homocysteine, lactic acid, ammonia, parathyroid hormone, pyruvate, and rennin testing require collection and delivery on an ice slurry to prevent analyte deterioration. Light-sensitive analytes, like bilirubin; folate; vitamin A, B6, and B12; porphyrins, require the specimens need to be protected from light. Specimens for measuring cold agglutinins or cryoglobulins should be collected in tubes that are warmed to 37 °C via warm water cup or minibath and contain no additives or gel. The specimen should be transported, delivered, and maintained at 37 °C, even during specimen processing in the laboratory. Failure to keep the sample warm will produce falsely decreased results. In larger health systems, where specimens may be transported to a core laboratory facility or onto a referral laboratory, specimens may need to be transported in refrigerated conditions or even frozen. It is always good practice to call the laboratory or consult the laboratory test directory and specimen collection manual for specific handling and transport instructions.

Minimizing Preanalytical Issues and Maximizing Specimen Quality

A blood sample represents a snapshot of a patient's condition. If specimen quality is compromised—"garbage in"—then a patient diagnosis may be delayed, mismanaged, or missed entirely—"garbage out". This chapter reviewed preanalytical issues commonly observed and experienced in healthcare and emphasizes three things:

- Patients and physicians depend on quality laboratory results for early diagnosis of disease and prevention of complications, which has great relevance in limiting hospitalizations.
- Quality specimens and communication of any irregularities with specimen integrity are the responsibility of the entire healthcare team. Steps should be taken to minimize preanalytical variables that may negatively affect clinical laboratory results.
- Quality laboratory data begins with quality specimens—quality in equals quality out.

References

American Diabetes Association. (2016). 2. Classification and diagnosis of diabetes. *Diabetes Care, 39*(Suppl 1), S13–S22. PMID: 26696675.

Baer, D. M., Ernst, D. J., Willeford, S. I., & Gambino, R. (2006). Investigating elevated potassium values. *MLO: Medical Laboratory Observer, 38*(11), 24. 26, 30-1. PMID: 17225675.

Bohn, D., & Daneman, D. (2002). Diabetic ketoacidosis and cerebral edema. *Current Opinion in Pediatrics, 14*(3), 287–291. PMID: 12011666.

Bush, V., & Cohen, R. (2003). The evolution of evacuated blood collection tubes. *Laboratory Medicine, 34*, 304–310.

Carmichael, J. D., Bonert, V. S., Mirocha, J. M., & Melmed, S. (2009). The utility of oral glucose tolerance testing for diagnosis and assessment of treatment outcomes in 166 patients with acromegaly. *The Journal of Clinical Endocrinology and Metabolism, 94*(2), 523–527. PMID: 19033371.

Chan, A. Y., Swaminathan, R., & Cockram, C. S. (1989). Effectiveness of sodium fluoride as a preservative of glucose in blood. *Clinical Chemistry, 35*(2), 315–317. PMID: 2914384.

CLSI. (2007). Procedures for the collection of diagnostic blood specimens by venipuncture; Approved Standard. CLSI document GP41-A6 (6th ed.). Wayne, PA: Clinical and Laboratory Standards Institute.

CLSI. (2010a). *Procedures for the handling and processing of blood specimens for common laboratory tests; Approved guideline. CLSI document GP44-A4 (4th ed.).* Wayne, PA: Clinical and Laboratory Standards Institute.

CLSI. (2010b). *Validation and verification of tubes for venous and capillary blood specimen collection; Approved guideline. CLSI document GP34-A.* Wayne, PA: Clinical and Laboratory Standards Institute.

Connor, C. S., Hermreck, A. S., & Thomas, J. H. (1988). Pitfalls in the diagnosis of pheochromocytoma. *The American Surgeon, 54*(10), 634–636. PMID: 3178050.

Cooper, G. R., Myers, G. L., Smith, S. J., & Sampson, E. J. (1988). Standardization of lipid, lipoprotein, and apolipoprotein measurements. *Clinical Chemistry, 34*(8B), B95–B105. PMID: 3042206.

Don, B. R., Sebastian, A., Cheitlin, M., Christiansen, M., & Schambelan, M. (1990). Pseudohyperkalemia caused by fist clenching during phlebotomy. *The New England Journal of Medicine, 322*(18), 1290–1292. PMID: 2325722.

Edge, J. A., Hawkins, M. M., Winter, D. L., & Dunger, D. B. (2001). The risk and outcome of cerebral oedema developing during diabetic ketoacidosis. *Archives of Disease in Childhood, 85*, 16. PMID: 11420189.

Elston, M. S., Sehgal, S., Du Toit, S., Yarndley, T., & Conaglen, J. V. (2016). Factitious Graves' disease due to biotin immunoassay interference—A case and review of the literature. *The Journal of Clinical Endocrinology and Metabolism, 101*(9), 3251–3255.

Ernst, D. J. (2005). *Applied phlebotomy, xviii* (p. 283). Baltimore, MD: Lippincott Williams & Wilkins.

Glaser, N., Barnett, P., McCaslin, I., Nelson, D., Trainor, J., Louie, J., et al. (2001). Risk factors for cerebral edema in children with diabetic ketoacidosis. The Pediatric Emergency Medicine Collaborative Research Committee of the American Academy of Pediatrics. *The New England Journal of Medicine, 344*(4), 264–269. PMID: 11172153.

Guder, W. G., Narayanan, S., Hermann Wisser, H., & Zawta, B. (2007). *Samples: From the patient to the laboratory: The impact of preanalytical variables on the quality of laboratory results* (3rd ed.). New York: Wiley. ISBN 9783527309818.

Head, K. A. (2006). Peripheral neuropathy: Pathogenic mechanisms and alternative therapies. *Alternative Medicine Review, 11*(4), 294–329. PMID: 17176168.

Jackson, S. L., Safo, S. E., Staimez, L. R., Olson, D. E., Narayan, K. M., Long, Q., et al. (2016). Glucose challenge test screening for prediabetes and early diabetes. *Diabetic Medicine*. doi:10.1111/dme.13270. PMID: 27727467.

Kummer, S., Hermsen, D., & Distelmaier, F. (2016). Biotin treatment mimicking Graves' disease. *The New England Journal of Medicine, 375*(7), 704–706. PMID: 27532849.

Lundberg, G. D. (1981). Acting on significant laboratory results. *JAMA, 245*(17), 1762–1763. PMID: 7218491.

Lundberg, G. D. (1999). How clinicians should use the diagnostic laboratory in a changing medical world. *Clinica Chimica Acta, 280*(1–2), 3–11. PMID: 10090519.

Meyer, R. D. (1986). Risk factors and comparisons of clinical nephrotoxicity of aminoglycosides. *The American Journal of Medicine, 80*(6B), 119–125. PMID: 3524214.

Narayanan, S. (2000). The preanalytic phase. An important component of laboratory medicine. *American Journal of Clinical Pathology, 113*(3), 429–452. PMID: 10705825.

Statland, B. E., & Winkel, P. (1977). Effects of preanalytical factors on the intraindividual variation of analytes in the blood of healthy subjects: Consideration of preparation of the subject and time of venipuncture. *CRC Critical Reviews in Clinical Laboratory Sciences, 8*(2), 105–144. PMID: 334466.

Stuart, K., Field, A., Raju, J., & Ramachandran, S. (2013). Postprandial reactive hypoglycaemia: Varying presentation patterns on extended glucose tolerance tests and possible therapeutic approaches. *Case Reports in Medicine, 2013*, 273957. PMID: 23424590.

Stumvoll, M., Mitrakou, A., Pimenta, W., Jenssen, T., Yki-Järvinen, H., Van Haeften, T., et al. (2000). Use of the oral glucose tolerance test to assess insulin release and insulin sensitivity. *Diabetes Care, 23*(3), 295–301. PMID: 10868854.

Wolfsdorf, J., Glaser, N., Sperling, M. A., & American Diabetes Association. (2006). Diabetic ketoacidosis in infants, children, and adolescents: A consensus statement from the American Diabetes Association. *Diabetes Care, 29*, 1150. PMID: 16644656.

Yatzidis, H., Koutsicos, D., Agroyannis, B., Papastephanidis, C., Francos-Plemenos, M., & Delatola, Z. (1984). Biotin in the management of uremic neurologic disorders. *Nephron, 36*(3), 183–186. PMID: 6322032.

Young, D. (2012). In C. A. Burtis, E. R. Ashwood, & D. E. Bruns (Eds.), *Tietz textbook of clinical chemistry and molecular diagnostics*, Chapter 6 (5th ed., pp. 119–144). London: Elsevier. ISBN 1455759422.

Yusof, A., Leithauser, R. M., Roth, H. J., Finkernagel, H., Wilson, M. T., & Beneke, R. (2007). Exercise-induced hemolysis is caused by protein modification and most evident during the early phase of an ultraendurance race. *Journal of Applied Physiology, 102*(2), 582–586. PMID: 17284654.

Zheng, Y., Schachern, P. A., Sone, M., & Papapella, M. M. (2001). Aminoglycoside ototoxicity. *Otology & Neurotology, 22*(2), 266–268. PMID: 11300281.

Chapter 4
Where the Lab Tests Are Performed: Analytical Issues in the Clinical Laboratory

James Miller

Congratulations! You have found where the lab tests are actually performed, now what kind of laboratory is it? There are several major types of clinical laboratories each with many similarities and differences: hospital laboratories, referral laboratories, true reference laboratories, outpatient clinic laboratories, and physician office laboratories. This chapter focuses on a typical medium-sized, tertiary care hospital laboratory. The first section of this chapter will describe how laboratories are physically organized, the types of equipment you are likely to see, and the people you may see running the analyzers and performing other functions within the laboratory. The second section will give a brief overview of automation and the analytical techniques used in the testing of biological specimens. The final section of this chapter will explain the analytical work that is "behind the scenes," that is, the quality assurance work that may not generate results on patient samples but is necessary to establish and maintain the quality of the analytical results.

Section One: Lab Equipment, Personnel

This section of the chapter describes the various sections of the laboratory and how they are physically and functionally organized and staffed.

J. Miller (✉)
Department of Pathology and Laboratory Medicine,
University of Louisville, Louisville, KY 40292, USA
e-mail: cmccudde@uottawa.ca

R. Molinaro et al. (eds.), *Clinical Core Laboratory Testing*,
DOI 10.1007/978-1-4899-7794-6_4

Hematology, Coagulation, and Urinalysis

Hematology deals with blood cell counting, including leukocytes, erythrocytes, and platelets, as well as the differential count of the leukocytes and morphology of the erythrocytes. To a large degree most of this work can be conducted by the automated hematology analyzer but the hematology laboratory can also perform the cell count manually. Most hematology analyzers today are based on flow cytometry. Various abnormal results from the analyzer may be programed to order preparation of a Wright-stained blood smear by the automated slide maker for manual review of the differential leukocyte count and/or the erythrocyte morphology. Hematology testing also includes the erythrocyte sedimentation rate, testing for erythrocyte sickling, staining for immature erythrocytes (reticulocytes) and other erythrocyte inclusion bodies, and detection and measurement of carboxyhemoglobin from carbon monoxide exposure. The hematology area also typically performs bone marrow evaluations and manual cell counts and microscopic evaluations on cerebral spinal fluid, serous fluids, and synovial fluid. Hemoglobin electrophoresis can be performed in hematology or within chemistry. More recently many hematology sections have added the enumeration of "lamellar bodies," or particles of phospholipids, as an aid to the evaluation of fetal lung maturity. This type of testing, traditionally done in chemistry, can be done within the platelet channel of automated hematology analyzers.

The coagulation laboratory is frequently combined with hematology, although in larger hospitals it could function independently. Even the smallest hospital labs perform prothrombin times (PT/INR) and activated partial thromboplastin time (aPTT). Coagulation laboratories also often perform PT and aPTT mixing studies, thrombin times, assays for specific coagulation factor deficiencies, von Willebrand factor assays, and platelet aggregation or platelet function tests.

Urinalysis is often conducted within the hematology section of the laboratory. Traditional urinalysis consisted of chemical measurement of several urinary constituents by dipstick (specific gravity, pH, leukocytes, nitrite, protein, glucose, ketones, urobilinogen, bilirubin, and blood) and a microscopic evaluation indicating the presence of cells, casts, and crystals, as well as any bacteria or other microorganism seen. The entire dipstick and microscopic analysis can now be accomplished by automated urinalysis instruments although identification of microscopic components presents the greatest challenge for automated analysis. However, images of the particles and cells identified are archived which can then be reviewed for accuracy by the technologist.

Clinical Chemistry and Toxicology

A wide variety of tests are performed in the clinical chemistry section of the lab. These include metabolites such as glucose, urea, creatinine, lactic acid, and ammonia; electrolytes, such as sodium, potassium, chloride, bicarbonate, calcium, magnesium, and phosphate; serum iron and iron binding capacity; lipids and lipoproteins, such as cholesterol, triglyceride, LDL-cholesterol, and HDL-cholesterol; and enzymes like alkaline phosphatase (Alp), aspartate aminotransferase (AST), alanine aminotransferase (ALT), creatine kinase (CK), and others. All of these are performed on automated general chemistry instruments.

Immunoassays

In the days of radioimmunoassays (RIA) immunoassays were traditionally done in a separate section of the laboratory, though usually considered part of the chemistry section. This is because of the special precautions required and specialized techniques used in this area. Since this area used antibodies as reagents, some laboratories called this section "immunology." During the 1990s non-isotopic immunoassay technologies became available in which enzymes and fluorescent and chemiluminescent compounds replaced radioactive atoms (most commonly ^{125}I) as labels in immunoassays. RIAs were largely discontinued as non-isotopic methods were introduced due to the regulatory requirements, health hazards of radiation, and the special requirements for disposal of radioactive waste. Immunoassays are now easily automated and widely used such that they are frequently integrated into every workflow within chemistry and often within other laboratory sections as well.

Immunoassay testing includes: hormones, such as thyroxine (T4), triiodothyronine (T3), TSH, cortisol, LH, FSH, prolactin, and hCG; hepatitis testing, like hepatitis B surface antigen; tumor markers such as alpha-fetoprotein (AFP), prostate specific antigen (PSA), and carcinoembryonic antigen (CEA); therapeutic drug monitoring, such as digoxin, vancomycin, and phenytoin; drugs of abuse screening for opiates, barbiturates, amphetamines, etc.; and major serum proteins, like prealbumin, haptoglobin, complement factors 3 and 4, and immunoglobulins G, A, and M. Many more immunoassay tests are available and which specific tests are performed by a laboratory is largely dependent upon its size and specialty expertise.

Therapeutic Drug Monitoring (TDM) and Toxicology

In most laboratories this testing is part of the chemistry section. As mentioned above it includes automated immunoassays for therapeutic drugs that need to be monitored because of toxicity risk such as antiepileptics (phenytoin, phenobarbital, carbamazepine, valproic acid); antibiotics (gentamycin, tobramycin, vancomycin); tricyclic antidepressants as a group or individually (amitriptyline, desipramine, doxepin, imipramine, nortriptyline); and a variety of other drugs, such as methotrexate, tacrolimus, cyclosporine, and digoxin. One of the few drugs not measured by immunoassay is lithium.

Drug of Abuse Testing (DAT), also called as Drugs of Abuse in Urine (DAU), includes testing for the recent use of illicit drugs (e.g., cocaine and phencyclidine), addictive therapeutic drugs (e.g., benzodiazepines, opiates, amphetamines), and agents used in drug withdrawal and treatment programs which are often diverted from their intended user (e.g., buprenorphine and methadone). These are assessed in urine by qualitative, drug class related immunoassays and are positive for a variable amount of time after use depending on the drug. It is important to realize that this type of testing is considered only a presumptive screen. This is often adequate to assist the emergency department in treating a patient, but there are times that a definitive, confirmatory assay is needed. This is most often done using mass spectrometry. An increasing number of labs have this technology for drug confirmation but this testing is a relatively expensive send out test.

Testing for volatile toxicants such as ethanol, methanol, isopropanol, acetone, and ethylene glycol is also commonly done in the toxicology lab by gas chromatography. Serum (and urine) osmolality is frequently performed in the toxicology section as well. Although osmolality is often used to assesses water balance or monitor mannitol therapy, the osmolar gap (difference between the measured osmolality and osmolality calculated from the sodium, glucose, and urea concentrations) can be useful for predicting the presence of volatile toxicants.

Most laboratories perform electrophoresis for serum proteins, urine proteins, and abnormal hemoglobin. In addition, many electrophoresis laboratories perform immunofixation to identify monoclonal immunoglobulins in serum and urine and the typing of positive cryoglobulins according to whether they include polyclonal or monoclonal antibodies, or both. Some larger laboratories may perform other forms of electrophoresis, such as lipoprotein electrophoresis, alkaline phosphatase isoenzymes, and isoelectric focusing for the CSF oligoclonal immunoglobulins of multiple sclerosis and other CNS diseases.

Blood Gas Assessment

Blood gas measurements may be done by the clinical laboratory or by respiratory therapists in the hospital. If done by the laboratory, it is usually done in the chemistry section.

Microbiology, Serology, Immunology, and Molecular Diagnostics

The microbiology section of the laboratory performs a vast array of testing. The culture, identification, and antibiotic sensitivity testing for bacteria are what we think of first and much of this work is automated now. However, there are many other kinds of pathologic microorganisms tested for by one technique or another, including fungi, yeasts and molds, parasites, and viruses. Technologists in microbiology perform the Gram stain and a variety of other stains directly on samples or on colonies grown in culture. They may use latex agglutination or immunofluorescent labeling to detect antibodies to various microbial antigens or detect the antigens themselves. This testing is categorized as serology and may be done in a dedicated section, but commonly performed in microbiology. Hepatitis serology (testing for hepatitis B surface antigen, hepatitis B core antibody, hepatitis An IgM antibody, and hepatitis C antibody) is generally the exception and commonly performed in chemistry/immunoassay areas because the hepatitis serology tests were the first to be automated and a time where automated testing resided only in chemistry.

Molecular Diagnostics

There are many applications of DNA or RNA testing in the clinical laboratory but the two most common today are related to identification, speciation, and drug susceptibility evaluation for microorganisms and pharmacogenetics. Although the various techniques used in molecular diagnostics are considered chemistry, the applications to microbiology are commonly performed in an enclosed area of the microbiology section. These include nucleic acid amplification techniques (NAAT) such as polymerase chain reaction (PCR) and restriction fragment length polymorphism (RFLP) detection, Western blots, characterization of ribosomal RNA, and viral load estimates. Pharmacogenetics may be considered a subspecialty of TDM but usually performed in a separate laboratory. Genetic testing for heritable diseases is usually sent out to referral laboratories. The technology used in molecular diagnostics is similar and although it would be cost effective to combine all testing into a single section, duplication of instrumentation for different applications of molecular diagnostics remains a common practice.

Laboratory Staff

The laboratory staff members that analyze patient specimens are either Clinical Laboratory Scientists (CLS), also known as Medical Technologists (MT), or Medical Laboratory Technicians (MLT). You may also see section supervisors or technical coordinators, who may occasionally work at the bench analyzing patient specimens. Supervisors usually report to laboratory administrators or to an administrative director depending on the size of the laboratory. These administrators can be MTs who have years of experience and may often have master's degrees in medical technology or business administration. You may also see doctorate level medical directors and technical directors and pathologists.

As mentioned above, MTs and MLTs are professional who do the analytical work, performing diagnostic analyses on specimens, including blood, urine, stools, sputum, and cerebrospinal, pleural, pericardial, peritoneal, and synovial fluids. MTs have earned a bachelor's or a master's degree in medical technology or an allied science such as biochemistry or microbiology. MLTs or technician-level lab workers have earned a 2-year associate degree in medical technology. Both MTs and MLTs analyze specimens, verify results based on a variety of instrument function indicators, and solve problems when results of individual tests or panels of tests do not fit the clinical picture. However, a technologist-level employee (MT) is required to be present when technicians are performing laboratory procedures. In larger laboratories, there are general supervisors available 24/7.

Laboratory professionals perform routine and specialized tests on the various specimens listed above using sophisticated laboratory equipment to gather data used to determine the absence, presence, extent, and cause of diseases. They troubleshoot equipment and procedures. Accuracy and quality is paramount in their work and they will frequently recheck their work whenever doubts arise regarding the validity of results reported. In collaboration with supervisors, administrators, and directors, technologists help make decisions about the optimal methods to use based upon sensitivity, specificity, ease of performance, and economy. Technologists also work with their superiors to establish effective quality control programs and oversee proficiency testing.

The MT and MLT must be organized, be able to multitask while paying attention to details, and work well under pressure. They must be adept at working with computers and a variety of computer applications. They must understand the theory behind the performance of specific tests and the significance of those results. Much of the work is performed on automated equipment and it is important for laboratory workers to understand what is going on inside the "black box" of these automated instruments.

Section Two: Automation and Technical Principles

Most clinical laboratories, especially mid-sized to large hospital laboratories, are highly automated. During the last 50 years or more a number of trends have increased the need for automation and its availability. These trends include among others: increasing numbers of laboratory tests and the consequent increasing number of tests ordered; desire for shorter turnaround times and the availability of testing around the clock; computerization of the analytical steps including the development of smaller and faster hardware and more sophisticated software; and a growing staff shortage.

The automated systems are highly technical instruments that require intelligent, well-educated, experienced technologists to properly run and understand them. The technologist must understand a large part of what is going on internally and be able to recognize when something is performing as expected.

Today there are automated systems for specimen processing (preanalytical), most types of testing (analytical), specimen storage (post-analytical), and tracks for automating the movement of samples from one stage to another. Automation has improved patient care in a number of ways. Automation has improved the integrity of samples by providing improved identity thru bar coding and primary tube sampling obviating the need to manually transfer samples from collection tubes to sample cups. Automation has also improved the accuracy and precision of testing. In addition, it has decreased manual handling and the risks of disease transmission to technologists. Finally, automation has improved efficiency, not only by automating previously manual work but also because automated instruments require less calibration and less maintenance and have less downtime.

Chemistry was the first to automate analyses in the 1950s and remains the most highly automated section of the laboratory. Initially, it was primarily the high volume general chemistry analytes that were automated. Separate immunoassay analyzers were later developed and today they are incorporated into the same instrument.

Automation in hematology was not far behind chemistry and is the second most automated section of the lab today. The development of automated cell counting beginning in the 1960s has greatly increased the accuracy and precision of counting red blood cells, platelets, white blood cells, and the differential count of the various types of white blood cells. The development of flow cytometry, the enumeration and characterization of a single file flow of blood cells, has provided further improvements to routine hematology and, when combined with fluorescently tagged white blood cell surface antibodies, has revolutionized the diagnosis of hematopoietic neoplasms.

All steps in the analytical process have been automated including sample aspiration, mixing sample and reagents, incubation, monitoring the reaction as it progresses or after a period of time, and calculations of the result. There are a few especially important points in the process. Most if not all automated analyzers have bar code readers and can sample directly from the primary sample collection tube. Whenever possible, that is preferred to eliminate the need to transfer an aliquot of the sample from one container to another which is a point at which an identification error can be made. Of course, if the aliquot is prepared automatically, a labeling error is unlikely. However, most analyzers require the sample tube stopper to be removed. Those steps possess a health hazard if done manually, but of course, the stopper removal can be done automatically. A few analyzers can sample directly thru the stopper, which eliminates the health hazard and the need to restopper the tube when all analyses are complete.

Some analyzers have a sample probe that aspirates the required volume of sample and place it in the reaction cup. Afterwards the sample probe is washed thoroughly in preparation for the next sample. Regardless of how thoroughly the probe is washed, there is the possibility of carryover and carryover must be assessed during method evaluation and periodically after that. Analyzers that use separate tips for each sample cannot have sample to sample carryover, so carryover does not have to be checked during evaluation or periodically afterwards.

Most analyzers have a mechanism to detect clots or particulate matter in the sample as well as liquid level sensors that can detect if there is an insufficient volume of sample for the analyses requested. However, some perform better than others. After detecting a short sample or a clot, some analyzers discard the aliquot that was picked up by the sample probe; others replace that aliquot into the tube. The latter option is especially valuable when small volume pediatric samples are analyzed and the loss of some of the sample may not leave sufficient volume in the tube.

Automated instruments have a variety of ways to handle reagents from bulk reagents in large containers to individual unit test reagents sufficient for one assay; some reagents require preparation by the technologist or the preparation may be automated. Depending on the analyzer's requirements for reagents, more or less storage space will be required for sufficient reagent between shipments. Even if no preparation of reagents is required, some reagents must be allowed to equilibrate to room temperature after obtaining them from a refrigerator or freezer. If not handled according to proper procedure, the reagent may be damaged. Automated analyzers may have the capability to keep track of how much reagent has been used over time, which can help automate restocking of reagents and other expendable supplies.

There are some disadvantages of automation. As mentioned above, technologists need to understand what is going on inside the "black box" of automated instruments. The automated methods for some analyses are not as accurate as the previous manual methods and should be considered as screening test followed up by the more accurate tests when appropriate. The decision for a laboratory to purchase one or another automated system is largely based on factors other than the quality of the assays on the instrument. It is not feasible for most laboratories to have several different automated analyzers for the same menu of tests. Finally, the rapid and relatively inexpensive methods on automated instruments tend to increase the frequency and variety of tests ordered often leading to an accumulation of irrelevant, unnecessary data. Not only does this tendency directly increase the cost of health care, but it inevitably leads to false positive results that must be followed up by more expensive testing.

Technical Principles

Spectrophotometry

Spectroscopy has been one of the major technologies in the clinical laboratory since its inception. Spectroscopy is useful for both qualitative and quantitative methods of measuring analytes in blood, urine, and other body fluids. The fundamental principles were studied by Lambert and Beer, who developed the well-known Beer–Lambert law or simply Beer's law.

Beer's law states that if monochromatic electromagnetic radiation (Po) is directed toward a container of a solution of an absorbing species, the amount of light transmitted is equal to the ratio of the light exiting the solution to the light entering the solution. This can be converted to the more useful expression in terms of absorbance (A), which is proportional to the concentration of thee absorbing species, thus, $A = abc$, where a is the molar absorptivity of the species in L mol^{-1} cm^{-1}, b is path length in cm, and c is concentration in mol/L.

This relationship is of critical importance in all assay reactions that depend on the development or disappearance of a "color," whether that "color" can be seen, that is, it is in the visible spectrum, or it is in the ultraviolet range of the spectrum. The vast majority of assays in chemistry and some assays in other sections of the lab use this analytical principle. A graph of absorbance versus concentration shows a linear plot, up to a point, with a y-intercept of zero and a slope of "*ab*." Obviously the relationship between absorbance and concentration must be in the linear range for accurate concentrations to be measured. High absorbance values and other factors, such as reagent lot changes and ambient temperature, may cause deviations from Beer's law and determining the linear range of an analysis is a major component of method validation and periodic quality assurance. It can be inferred from the principles of spectrophotometry that chemicals or compounds that absorb light may interfere with analytical testing (see Interferences, later in the chapter).

Sometimes it is more convenient to measure the absorbance in reflected light rather than transmitted light and this form of spectrophotometry is called reflectometry. This technique is used in urine dipstick analysis and dry slide chemical technology, where the reagents are imbedded in pads or layers of film (Table 4.1).

Fluorometry

Fluorescent compounds will absorb light at a specific wavelength and an electron becomes excited to a higher energy level. Depending on the compound, after a period of nanoseconds to microseconds in the excited state, during which some vibrational energy is lost, the electron returns to the ground state with the release of a photon of longer wavelength than had been absorbed. Fluorometry is widely used

Table 4.1 Common laboratory tests and measurement principles

Test	Principle	Relative turnaround time
Sodium, potassium, chloride	Electrochemistry	Very fast
AST, ALT, ALP	Spectrophotometry	Fast
TSH, hCG, troponin	Fluorometry or turbidimetry	Moderate
Osmolality	Osmometry	Fast
Complete blood count	Flow cytometry	Fast
Porphyrins, metanephrines, ethylene glycol	Chromatography	Slow[a]
Serum proteins, oligoclonal bands	Electrophoresis	Slow[a]
B2-microglobulin, cystatin C, immunoglobulin subclasses	Nephelometry	Moderate
ACTH, renin	Radioimmunoassay	Very slow[a]
Tacrolimus, cyclosporine	Mass spectrometry	Moderate-slow
Lead, mercury	Atomic absorbance spectroscopy	Slow[a]

[a]These tests are often run in batches and are usually run <7 days/week

because of its high sensitivity, which in turn is due to the high signal-to-noise ratio of this technique. Fluorometry is also of high specificity because of the combination of a particular excitation wavelength and a particular emission wavelength of a compound.

There are several useful variations on simple fluorometry, often used in immunoassays. In time-resolved fluorescence the energy of the excited state is transferred to a species with long-lived fluorescence such as a chelate of europium. This technique increases the sensitivity and specificity by minimizing background signal and some interferences. In chemiluminescence the fluorescent compound is excited chemically or even by an electrical signal rather than by light. In fluorescent polarization immunoassays (FPIA) a fluorescently labeled tracer competes with the analyte of interest for binding to the antibody. After incubation the label (usually fluorescein) is excited with plane-polarized light (494 nm for fluorescein). Unbound label will rotate rapidly, losing its polarization, while antibody bound label will rotate relatively slowly, maintaining more of its polarization. At a 90° angle the amount of emitted light (521 nm for fluorescein) that is still plane-polarized is measured and this is proportional to the amount of label bound to the antibody and can be used to quantitate the analyte in the unknown sample.

Nephelometry and Turbidimetry

Nephelometry and turbidimetry are both techniques that measure the light scattered by large particles (such as antigen–antibody immune complexes). Nephelometry actually measures the scattered light at some angle (15–90°) to the incident light, while turbidimetry measures the decrease in light transmitted due to the scattered light. Both techniques are used predominantly in immunoassays, frequently for the so-called major serum proteins, like prealbumin, haptoglobin, transferrin, immunoglobulins, and others. Turbidimetry has become quite common because it is much easier to implement on automated instruments than is nephelometry. In addition, a large variety of turbidimetric formats have been developed by attaching antigens or antibodies to microparticles.

Osmometry

In osmometry the osmolality of serum, plasma, or urine is measured. Osmotically active particles, such as glucose, urea, and electrolytes, increase the osmolality of the fluid. Osmometers are based on measuring the boiling point or freezing point of the sample, which are proportional to the osmolality. Freezing-point depression osmometry is the most useful in the clinical laboratory because it can detect volatile toxicants such as ethanol, methanol, isopropanol, acetone, and ethylene glycol.

Flow Cytometry

In a flow cytometer, particles such as blood cells are forced to flow single file past a laser light source. Each particle produces a characteristic pattern of scattering and size that is measured by detectors. This is sufficient for counting and differentiating leukocytes, erythrocytes, and platelets in hematology. Detailed analysis of types of leukocytes, mainly lymphocytes, can be accomplished by preincubating the sample with fluorescently labeled antibodies to surface antigens or even intracellular components.

Electrochemistry

Electrochemistry involves measurement of the current or voltage generated by the activity of specific ions. Analytic techniques include potentiometry, coulometry, voltammetry, and amperometry. These techniques are used to measure sodium, potassium, ionized calcium, and other ions using ion-selective electrodes; and pH, pCO_2, and pO_2 in blood gas analyzers.

Electrophoresis and Densitometry

In electrophoresis charged compounds are separated according to charge. The greater the net charges of a compound, usually a protein, the faster it moves. The movement of individual proteins, and therefore the resolution of the mixture, can be adjusted by changes in the pH and ionic strength of the buffer, the voltage applied, and the time of separation. Various supports have been used, including cellulose acetate, agarose, and polyacrylamide gel. After the separation, the proteins usually are fixed and stained with a protein dye. If lipoproteins are being separated, the support may be stained with a general lipid dye or cholesterol-specific reaction. The stained support is typically scanned by a densitometer, which is similar to a spectrophotometer, but with the ability to move the support past the detector. This generates an electrophoretogram or graph of absorbance versus electrophoretic mobility. The densitometer also quantitates the total amount of protein in each peak or fraction. Electrophoresis with densitometric scans is commonly performed on small samples (about 1 μL) of serum and urine proteins and on whole blood hemolysates for detection of abnormal hemoglobins. In immunofixation electrophoresis, usually to detect monoclonal antibodies, the antibody classes are immunostained with anti-IgG, A, and M as well as anti-kappa and lambda. These are then interpreted by eye without densitometry. Although electrophoresis is a relatively labor-intensive technique, a good deal of the actual electrophoresis and densitometry has been automated.

Capillary Electrophoresis

Capillary electrophoresis is performed in a very narrow bore fused silica capillary. Otherwise the electrophoresis is similar to classical electrophoresis, except that detection occurs as the proteins, or other analytes, flow past the detector. Capillary electrophoresis uses a smaller sample size than classical electrophoresis (in the nL range), it is rapid providing short turnaround times in the range of 10 min, it has high resolution, and it is more easily automated. The utility and acceptance of capillary electrophoresis is growing.

Isoelectric Focusing

Isoelectric focusing is similar to electrophoresis, but with the inclusion of ampholytes in the buffer. Ampholytes are mixtures of small amphoteric compounds with a range of isoelectric points (pI). During the electrophoresis, the ampholytes migrate according to their pI, which generates a pH gradient in the gel. Each specific protein has a specific pI depending on the number of positively and negatively charged amino acid side chains in its structure. Each protein will migrate to its pI and will be focused there because any tendency to diffuse away will be counteracted by the drive to its pI. Isoelectric focusing is used when high resolution is required, for example, for detecting the oligoclonal immunoglobulins in multiple sclerosis or in phenotyping alpha-1-antitrypsin deficiency.

Chromatography

Chromatography is a versatile separation technique based on the partitioning of compounds between a stationary phase and a mobile phase. If under the conditions used a compound interacts more strongly with the stationary phase, the compound will migrate more slowly. Conversely, if a compound interacts more strongly with the mobile phase, the compound will migrate more quickly. The wide variety of types of chromatography is related to the many types of these phases available. The most common types of chromatography used in the clinical lab include gas liquid chromatography for separating volatile compounds like ethanol, methanol, isopropanol, and acetone; high-pressure liquid chromatography (HPLC) commonly used to separate compounds in preparation for mass spectrometry; and ion-exchange HPLC for separating hemoglobin variants and also useful for quantitating hemoglobin A_{1c}.

Mass Spectrometry

In mass spectrometry (MS) molecules are fragmented and ionized, after which powerful magnets attract the ions to different degrees based on their mass-to-charge ratio. The mass fragmentation pattern or mass spectrum allows the identification and quantitation of the molecules. Typically before a mixture of compounds is introduced to the mass spectrometer, it is separated into individual compounds by gas chromatography or HPLC, the latter being more common today. Also the mass spectrometers of today are usually tandem MS, in which specific ions from the first MS are directed into a second MS where further fragmentation occurs. This LC-MS/MS allows for very accurate and specific measurement of compounds. Many smaller labs do not have this technology yet, but the availability is growing rapidly. This technique is usually used for therapeutic drug monitoring of immunosuppressive drugs and confirmation of positive drug screens. Any small molecule that can be ionized is amenable analysis by mass spectrometry.

Other Techniques

Other common techniques are listed in the description of laboratory sections below. There are many other techniques used in some larger laboratories or specialty laboratories. A few are mentioned here. Some larger laboratories use atomic absorption spectroscopy (AAS) to measure metals, for example, lead, mercury, copper, and zinc, in serum or tissues. Radioimmunoassays have been largely replaced by non-isotopic methods, obviating the regulatory issue, health risks, and disposal problems related to radioactive compounds. However, a few assays still require the use of radioactive reagents. Gamma counters or scintillation counters are used for these assays. There are specialty labs that use nuclear magnetic resonance spectroscopy to analyze the lipoprotein particle size distribution.

In addition, there are a variety of more general techniques that laboratory technologists use, for example, the preparation and testing of high quality water. While most reagents are purchased in a ready-to-use form these days, some reagents or calibrators may still have to be prepared manually from high quality chemicals. When preparing these reagents or calibrators, the technologist will have to use an analytical balance and the proper size and type of laboratory glassware or plasticware, including pipets and volumetric flasks. Technologists in hematology may have to perform a Wright's Stain on a blood smear or know how to run the automated stainer. In microbiology techs must know how to perform a Gram's Stain. Other commonly used equipment include refrigerators, freezers, thermometers, incubators, water baths, centrifuges, autoclaves, microscopes, to name a few.

Quality Assurance

General

Quality is the degree to which a product or service satisfies the need for that product or service. In the laboratory we provide a service—results of laboratory tests. But our service is much more than just the results of laboratory tests. As described in the introduction, quality entails the results from the right tests at the right time on the right samples. The results must be sufficiently accurate and precise to provide useful clinical information to the ordering physician. Quality assurance is the variety of procedures and monitors put in place in the laboratory to assure all aspects of the quality of laboratory service.

From the definition above, one can see that there are many aspects to quality or laboratory results must meet the needs of physicians and patients. Much analytical work has been done behind the scenes before any patient results are produced, both before introducing a new test or change in a test methodology (method evaluation) and on an ongoing basis (Quality Control and Proficiency Testing). The testing done before introducing a new test or change in test methodology is called method validation or method evaluation. On a day-to-day and periodic basis quality control samples are run, proficiency samples are run, and maintenance is performed. This section will outline the key analytical aspects of quality assurance. These include the following consideration and procedures:

- What tests should be provided?
- What methods should be used for those tests?
- How good do the methods have to be to provide clinically useful information?
- Method evaluation
  - Precision
  - Accuracy
  - Linearity
  - Method comparison
  - Interpretative comment and reference ranges
  - Carryover
  - Sample type
  - Analyte stability
  - Interferences
- Quality control
- Proficiency testing

Choosing Tests and Methodologies

The breadth of a laboratory test menus is largely dependent on the variety of services and acuity of the hospital or clinic. For example, hospitals that specialize in transplantation surgeries will offer therapeutic drug monitoring for a wide variety of immunosuppressant medications. Basically, if there are a sufficient number of orders for a particular test, the laboratory is likely to offer that test in house. On the other hand, if the number of orders for a test is below about 10–15 per week, it would most likely not be cost effective to offer the test in house and the turnaround time would likely be shorter if the test is sent to a referral laboratory (referral laboratories and test menus are further discussed in Chaps. 2 and 3).

Once it has been decided that a test will be performed in house, the laboratory must perform evaluations to be sure that the assay is of sufficient quality (method evaluation) and then ensure that the good performance is maintained (quality control).

How good do the methods have to be to provide clinically useful information? To a laboratorian this question is equivalent to asking "How much error is tolerable?" Error is an unavoidable fact of measurement. In any form of measurement there are two sources of error: bias and imprecision. Bias is defined by how far the measurement is from the true value, where a low bias is synonymous with accuracy (Fig. 4.1). If we measure the same thing repeatedly, it is not uncommon to get slightly different results sometimes. Imprecision is a measure of how much variability there is in those measurements (Fig. 4.1). Imprecision is usually expressed in the statistical term "standard deviation" or SD. The average of those repeated measurements is the "mean." If the true value is X, then bias is mean $-X$ and that can be represented as a percentage by dividing by X and multiplying by 100 %, that is,

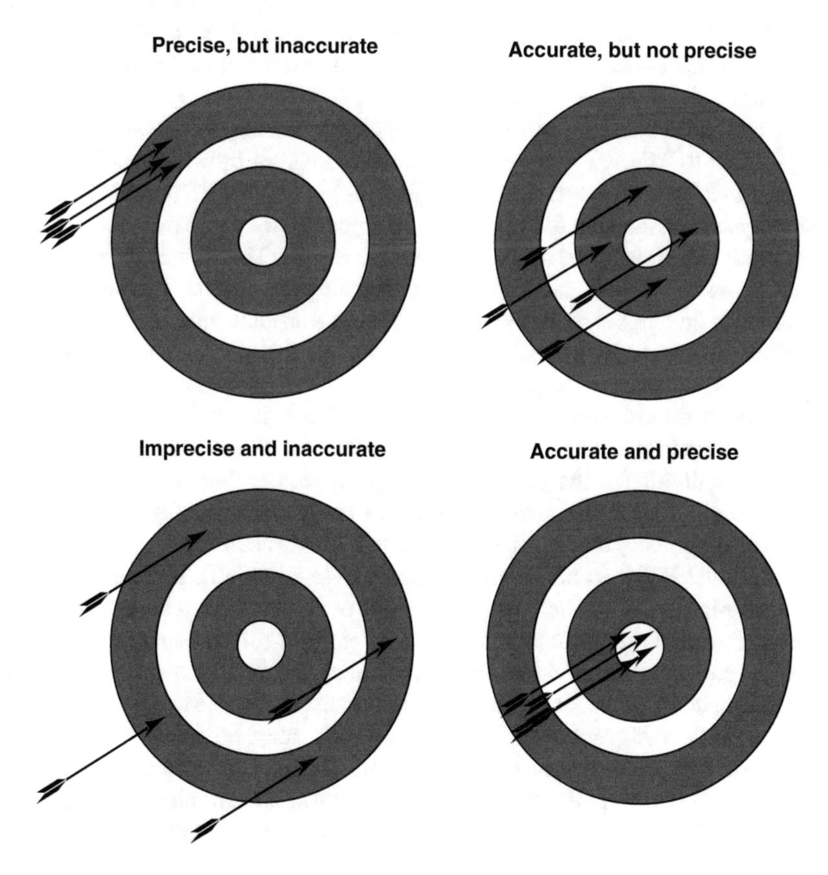

Fig. 4.1 Conceptual comparison of accuracy and precision

$$\text{Bias}(\%) = 100\,\% \times (\text{mean} - \text{X}) / \text{X}$$

SD is often expressed as a percent, in which case it is called the coefficient of variation (CV) defined as:

$$\text{CV}(\%) = 100\% \times \text{SD} / \text{mean}$$

The total error (TE) of a measurement is equal to the absolute value of the bias plus $2 \times$ imprecision. This formula for the total error can be expressed in terms of bias and SD or in terms of bias (%) and CV as follows.

$$\text{TE} = |\text{bias}| + 2\text{SD} \quad \text{or} \quad \text{TE}(\%) = |\text{Bias}(\%)| + 2\text{CV}(\%)$$

Now we can come back to the question "How good do the methods have to be to provide clinically useful information?" The answer to that question is called the Total Error Allowable or TEa. There are several methods to decide what the TEa should be and they don't all agree. A discussion of all of these methods is beyond the scope of this book, so the two most commonly used methods will be described briefly here.

One method is based on considering Proficiency Testing (PT) Evaluation Limits as equivalent to TEa. PT is a quality assurance program by which a PT provider, such as the College of American Pathologists (CAP) or other PT provider, sends the laboratory several (usually five) unknown specimens three times per year. The laboratory assays the specimens and sends back the results. The laboratory's results are judged by how close they are to the mean of their peers results. Each analyte (test substance) is given a set of limits (e.g., ±2 SD or an absolute number) and the laboratory's results should be within these limits. In this context, maximum error allowed (TEa) equates to the proficiency limits.

Another common method to decide on the TEa is to use biological variation (BV) (discussed further in Chap. 5). If an analyte is measured in a large number of healthy individuals, there will be a range of results that are consistent with health. This is called the healthy reference range or interindividual BV. It is useful for diagnosis, that is, if an individual's test result varies more than the interindividual BV, then his result is abnormal (low or high). If the analyte were measured repeatedly in the same individual over time, the results would vary somewhat around the individual's personal set point for that analyte. This variability is called intraindividual BV. Physicians frequently measure the same analyte repeatedly as occurs with inpatients or disease monitoring. For example, a patient with a high cholesterol concentration may be treated by diet and exercise or with medication to bring the cholesterol concentration down. The physician will periodically repeat the cholesterol test to monitor the success of the therapy. BV theory calculates the TEa from a combination of inter- and intraindividual BV.

Method Evaluation

With any new test or change in methodology there is the requirement for a method evaluation. Part of this validation relies on the using the above TEa to decide if a method is acceptable. The vast majority of methods used by hospitals have been cleared by the Food and Drug Administration (FDA). Such methods require that the vendor does extensive studies to demonstrate to the FDA that the method meets required standards (e.g., is similar to existing methods or has defined performance deemed acceptable by the FDA). Federal law, in the form of the Clinical Laboratory Improvement Amendments of 1988 (CLIA), dictate many aspects of laboratory quality assurance including method evaluation. Thus, for an unmodified, FDA-cleared or approve method, before reporting any patient test results, the laboratory must perform studies to show that accuracy, precision, and reportable range (linearity) are satisfactory in the environment of the laboratory, and verification that the manufacturer's suggested reference intervals (normal values) are appropriate for the laboratory's patient population. If the method is not FDA-cleared or has been modified, additional components of method evaluation are required, including testing analytical sensitivity (limit of detection), analytical specificity (interference and cross-reactivity studies), establishment, not just verification, of the reference interval, and any other performance characteristics required for test performance, e.g., specimen type and analyte stability. All of this method validation work must be documented prior to method implementation.

Imprecision

A typical study of imprecision includes assaying two or more levels (e.g., low and high concentrations) of control material in duplicate, twice per day, for 20 days. This experiment yields 80 measurements of each level of control. By analysis of variance, the within-run imprecision, the run-to-run imprecision, the day-to-day imprecision, and the total imprecision can be calculated. The total imprecision in terms of SD or CV is the most important component of variability and is what we will judge against our TEa. Recall the formula for TE.

$$\text{TE} = |\text{bias}| + 2\text{SD} \quad \text{or} \quad \text{TE}(\%) = |\text{Bias}(\%)| + 2\text{CV}(\%)$$

One can see that in order for TE to be less than the TEa, SD (or CV) must be less than or equal to approximately one-third of the TE (or TE, %). For example, the PT evaluation limit for glucose is 6 mg/dL or 10 %, depending on if the glucose concentration is below or above 60 mg/dL. Let's say we are evaluating a new glucose method and one of our control glucose concentrations has a mean of 90 mg/dL and an SD of 2.5 mg/dL. The glucose concentration is greater than 60 mg/dL, so we need to judge this imprecision based on 10 %. To convert the SD to CV (%) we divide the SD by the mean and multiply by 100 %. Thus,

$$\mathrm{CV}(\%) = 100\% \times \mathrm{SD} / \mathrm{mean} = \mathrm{CV}(\%) = 100\% \times 2.5\ \mathrm{mg/dL} / 90\ \mathrm{mg/dL} = 2.8\%$$

In this example, an imprecision of 2.8 % looks pretty good for glucose if we want to keep our TEa below 10 %, but an estimate of the bias is still required to be certain.

Accuracy

The goal of accuracy studies is to determine if the method is fit for use. Combined with imprecision estimates, the total error is compared against the allowable error (TEa). Continuing with the glucose example, suppose the mean of laboratory peers (X) is 92 mg/dL and the mean in a precision study was 90 mg/dL. Then the test lab averaged 2 mg/dL below the peers.

$$\mathrm{Bias}(\%) = 100\% \times (\mathrm{mean} - X) / X = 100\% \times (90\ \mathrm{mg/dL} - 92\ \mathrm{mg/dL}) / 92\ \mathrm{mg/dL} = \text{-}2.2\%$$

Combined with the earlier imprecision estimate, the total error is 7.8 %:

$$\mathrm{TE}(\%) = |\mathrm{Bias}(\%)| + 2\mathrm{CV}(\%) = \mathrm{TE}(\%) = |-2.2\%| + 2 \times 2.8\% = 7.8\%$$

This TE is less than the TEa of 10 % (from above based on proficiency testing), such that the method's total error is interpreted as acceptable. Alternatively, biological variation goals could be used (http://www.westgard.com/biodatabase1.htm). In either case, the decision to accept and implement a new method or test is ultimately made by the laboratory director. If a method does not appear to meet the clinical needs of the lab, then the manufacturer is contacted to help troubleshoot the method or instrumentation or another method is sought.

Method Comparison

The method comparison is simply a comparison of patient specimen results using the new assay method with the results using a comparison method, which could be a reference method or an existing method targeted for replacement. This data is graphed with the results of the comparison (reference or other) method on the *x*-axis and the new method results on the *y*-axis. If the two assays give essentially the same results, the graph will appear to be a straight line with a slope of 1.0 and a y-intercept of 0 (Fig. 4.2).

This experiment is often considered to be a measure of accuracy, but it is only a measure of accuracy if the comparison method is a reference method. In reality reference methods are not readily available to the typical laboratory and there is no reference method for many analytes. In practice the comparison method is the old method. If this test is a new one for the lab, chances are that previous requests for

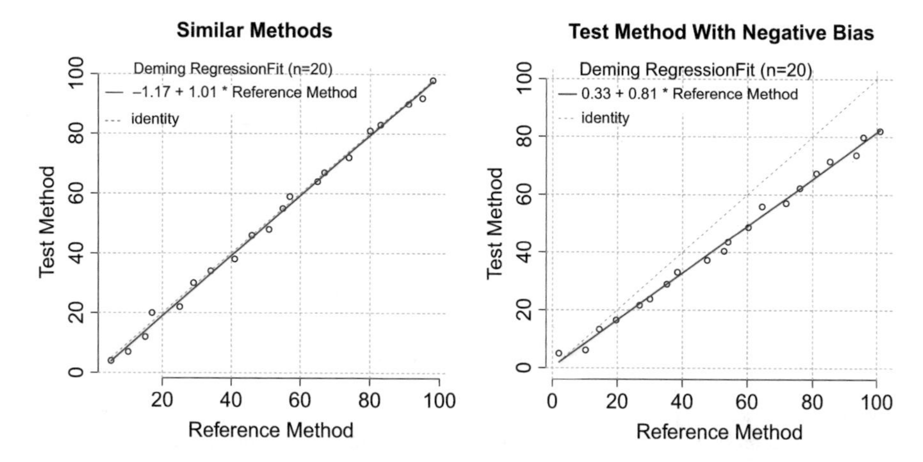

Fig. 4.2 Typical method comparison analysis graphs. The *left* panel shows a regression plot for a set of methods were the values are very similar (note the slope and intercept are near 1 and 0, respectively). The *right* panel shows two dissimilar methods. In this instance there is a proportional bias displaying a slope of much less than one; the intercept (constant bias estimate) remains close to zero

the test were sent to another lab, a referral lab. In that case, the results from the referral lab would serve as the comparison method and be graphed on the x-axis.

The method comparison experiment tells whether to expect somewhat different results with the new method, higher or lower than the old method and this information allows the laboratory to decide whether to write a memo to the physicians about the change in methodology. In addition, the method comparison experiment provides information about whether the manufacturer's suggested reference range may be valid. And finally, this experiment may help identify unusual problems if one or more sample does not compare as well as most samples do.

Reportable Range (Linearity)

The reportable range of an assay is the range between the lowest concentration we can measure and the highest concentration that is still within the linear range of the assay. Oddly, CLIA does not require us to validate the low end of the reportable range (but see limit of detection below) for an FDA-approved method, just the upper end. To do this we find a sample, patient sample, control or calibrator with the highest concentration of the analyte that may be available. We make multiple dilutions of this sample, so we have diluted sample across the analytical range. We graph the expected results against the measured results to determine the range of concentrations that are linear on the plot (Fig. 4.3). This gives us the upper end of the reportable range and we would accept the manufacturer's claimed limit of detection or limit of quantification as the low end of the analytical range.

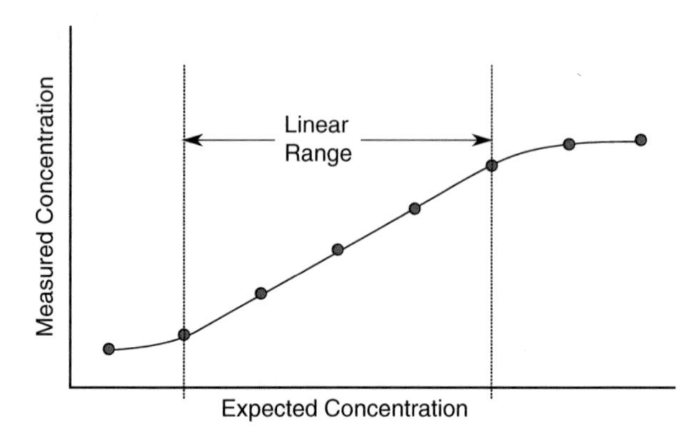

Fig. 4.3 The linear range of a method is the portion where the response is directly proportional to the concentration in a straight line. It is common for very high or low values to yield nonlinear responses, such that they cannot be measured accurately. High concentrations may be diluted for some analytes, but low values are typically reported as "less than" the number at which the response is linear

Reference Range Validation

For FDA-approved methods the lab is required to validate the manufacturer's suggested reference intervals (normal range). This is relatively easy to perform. At least 20 samples need to be assayed that were collected from relatively healthy volunteers, or others known to have normal results for the test under evaluation. If no more than 10 % of the results are outside the suggested normal range, that normal range is accepted as suitable for our population. If more than 10 % of the results are abnormal, the study can be repeated with another 20 samples. If there are still more than 10 % abnormal results, it is likely that there is either a method problem, or a new reference interval needs to be established. To establish a new reference interval, it is generally accepted that a minimum of 120 samples be collected from healthy volunteers. That can be quite difficult and expensive since and could end up being many more than this if there are gender differences or age-related differences. Methods developed in-house (a method that is not FDA approved), also require larger (more than 20 samples) reference interval studies.

In general, non-laboratorians are surprised at how few samples are used to establish the reference ranges that are so widely used to make clinical decisions. While the 120 sample minimum is based on some statistical sampling, consider the limitations of the 20 sample reference interval validation from a population standpoint. If there are regional population differences, such as latitude, diet, and race, between the reference and validation samples then it is likely that the reference interval will be of limited utility for clinical decisions. It is for this reason that laboratorians advise and physicians develop an intuition for relative change rather than rigorously applying 95 % cutoffs to patient data.

Other Method Evaluation Experiments

As mentioned above, methods that are not FDA-cleared or approved or has been modified must be evaluated more thoroughly. In addition to the testing above, the laboratory must test the analytical sensitivity (limit of detection), analytical specificity (interference studies), and establish rather than simply verify, of the reference intervals; other performance characteristics that may be required for test performance include specimen type (serum, plasma, type of anticoagulant, etc.) and analyte stability. Space does not allow a description of these additional experiments, but it is important to consider interferences, that may occasionally cause inaccurate results or prevent the lab from reporting any results.

Interferences

Sometimes patient samples contain interfering substance. These may be endogenous substances such as elevated bilirubin, elevated lipids, or interfering antibodies or they may be exogenous substances such as drugs or components of the patient's diet. An interference occurs when a specimen component other than the analyte itself alters the concentration measurement.

The major endogenous interferences include hemoglobin and other components released from damaged RBCs (hemolysis), turbidity most commonly caused by elevated triglyceride-rich lipoproteins (lipemia), and elevated bilirubin (icterus). These interferences impart an abnormal color to serum or plasma sample which can be assessed visually. However, most large chemistry analyzers today automatically check for hemolysis, turbidity, and icterus and flag the results with an H, T (or L), or I, respectively, if there is too much of these interferences.

Whole chapters or even whole books could be written about interferences. There is only space enough here to briefly describe some basic concepts in the area of interferences and a few of the major types of interferences.

Basic Concepts

As mentioned above, interference causes inaccuracy or bias in clinical laboratory results. The bias may be positive (falsely increasing the result) or negative (falsely decreasing the result). Both the direction (positive or negative) of interference and the magnitude of interference of a particular substance may vary in different assay methods. Recall from above that we work to keep the TE of our assays below the TEa. The added bias from interferences only becomes clinically significant if it causes the TE to exceed the TEa, so sometimes we can get by with a small amount of interference.

Interference in Immunoassays

Immunoassays are assays that use antibodies as reagents. Because immunoassays are a large and growing type of assay in the clinical laboratory, this section reviews several types of interference pertinent to immunoassays, namely cross-reactivity and interfering antibodies.

Cross-Reactivity

Cross-reactivity occurs when a compound in the specimen has a structure similar to the analyte and competes for binding to the reagent antibody. The most common source of cross-reactivity is drug metabolites interfering with the measurement of the parent drug. For example, a physician may want to measure the hormone cortisol in a patient that is taking prednisone. Prednisone and its metabolites such as prednisolone have structural similarity to cortisol and may cause falsely elevated cortisol results in some assays due to cross-reactivity.

Heterophile and Other Interfering Antibodies

Antibodies in patient's blood serum can sometimes react with reagent antibodies and cause interference. Most of the time such interference will be a false increase in result. Clinical Chemists differentiate two types of interfering patient antibodies, anti-animal antibodies and heterophile antibodies, but the differences between them are minor and their effects are essentially the same. Anti-animal antibodies are antibodies that the patient developed because of exposure to the particular animal. For example, dairy farmers may develop antibodies to bovine proteins and grain elevator workers may develop antibodies to mouse proteins. The antibody reagents in immunoassays are proteins and usually they are animal proteins, often mouse, sheep, or rabbit. So a patient that happens to have antibodies to mouse proteins may get a falsely elevated result in an immunoassay that uses mouse antibodies. This particular interfering antibody would be termed a "human anti-mouse antibody or HAMA." There are many reports of falsely elevated results due to HAMA. In many cases patients have developed HAMA after being treated with therapeutic antibodies developed in mice.

The other type of interfering antibody is called a heterophile antibody. The term heterophile antibody simply means that it is an antibody that reacts with proteins from another species, which sounds almost identical to anti-animal antibodies. The only difference is that in the case of heterophile antibodies, no specific animal exposure can be identified and the strength of the binding to the reagent antibodies is weaker than in the case of anti-animal antibodies. There are also many reports of heterophile antibody interference in the literature.

Although there are many reports of both HAMA and other anti-animal antibodies and heterophile antibodies in the literature, that does not mean they are common. In fact, falsely increase results probably happen only about one time in 10,000 assays, but most of the reports are in very common assays and especially those in which the typical result is very near zero. In assays like this, a slightly increased result that does not fit the clinical picture is relatively easy to recognize (e.g., elevated hCG). As with the hook effect, diagnostic assay manufacturers have developed better ways to block these interfering antibodies and we are also getting better at noticing when they occur.

Quality Control

Having completed the method evaluation and concluded that the assay method has suitable quality, it is time to put the assay into routine use. A major aspect of quality assurance is quality control. Quality control (QC) is the practice and processes whereby QC samples with known analyte concentrations are assayed and the results checked to be within acceptable limits before patient samples are assayed. The daily QC results are recorded and graphed to show the pattern of the distribution of results for each day. The mean of the control results is set as the center of the chart and markers for the acceptable variation from the mean (often 2 SD) lie above and below the mean. The mean and SD had been determined in the method evaluation studies. If the result on the control is outside of expected range, then corrective action is required. This corrective action may be as simple as repeating the analysis of the control or as complex as changing the reagents or recalibrating the assay. Only when the results on the control are within acceptable range can patient samples be analyzed.

Proficiency Testing

In addition to daily quality control samples, laboratories are required to participate in proficiency testing (PT) programs. There are several different organizations that provide proficiency challenges, but in essence they each send unknown or blind samples to participating laboratories for analysis. Labs submit their results to the provider within a short time frame and the program calculates the mean and other statistics for the data submitted and determines if the results from each lab are sufficiently close to the mean of peers or sometimes close enough to the true results based on assay by a reference method. Typically, a set of five unknown samples are sent three times per year. At least four out of the five specimens must be within the Evaluation Limit from the mean of peers or the reference value. Evaluation Limits for selected analytes are

Table 4.2 Proficiency testing evaluation limits

Analyte	Evaluation Limit
Sodium	4 mmol/L
Potassium	0.5 mmol/L
Chloride	5 %
Alkaline phosphatase	30 %
Calcium	1 mg/dL
Magnesium	25 %
Hemoglobin	7 %
Hematocrit	6 %
Leukocyte count	15 %

shown in the table. If the laboratory repeatedly fails this PT, the laboratory may be in jeopardy of having to discontinue a test and, if there is a problem with several tests, the laboratory may lose its certificate to do business (Table 4.2).

Summary

There is a substantial amount of work and analysis that goes into validating and maintaining the quality of each laboratory test. These are all components of the analytical phase of testing that are "behind the scenes" in that they do not produce patient results. However, this quality assurance work is absolutely necessary to ensure that the laboratory delivers quality patient results. This quality assurance work costs money, both in reagent usage and personnel time. Some say that quality assurance is expensive, but in reality it is the lack of quality that is expensive because the investment in good quality leads to improved patient outcomes, decreased need to repeat testing, shorter lengths of stay in the hospital, fewer readmissions for the same illness, and fewer medical malpractice suits. The combination of lower costs when the laboratory is in a state of quality saves much more money than the cost of the quality assurance.

Chapter 5
Post-analytical Issues in the Clinical Laboratory

Christopher R. McCudden and Monte S. Willis

Introduction

The final part of the laboratory testing process is known as the "Post-Analytical Phase." In this phase, lab results are communicated to physicians. While the mechanisms of communication vary from verbal to digital, the end goal remains the same: provide accurate, timely, informative results to the physician. Although physicians are on the receiving end of this process, it may remain a bit of a black box. Common questions include: Why is this test taking so long? Why did the reference interval change since last time? Why is the lab calling me? How do I interpret a modest change in a result over time?

The objectives of this chapter are to answer these questions with practical examples and highlight additional useful information that the laboratory can provide to help physicians interpret test results.

Why Is the Test I Ordered Taking So Long?

It is essential to have as much data as possible when taking care of patients to make the best decisions for their care. Critical to this process is having the results of laboratory tests available as soon as possible. This is even more important when patients

C.R. McCudden, Ph.D. (✉)
The Ottawa Hospital, General Campus, Pathology and Laboratory Medicine,
501 Smyth Road, Ottawa, ON, Canada, K1H 8L6
e-mail: cmccudde@uottawa.ca

M.S. Willis, M.D., Ph.D. (✉)
Department of Pathology and Laboratory Medicine, University of North Carolina Hospitals,
University of North Carolina, Chapel Hill, NC 27599, USA
e-mail: monte_willis@med.unc.edu

R. Molinaro et al. (eds.), *Clinical Core Laboratory Testing*,
DOI 10.1007/978-1-4899-7794-6_5

are acutely ill, in the emergency department, and/or the intensive care units. So it is not uncommon to wonder why physician-ordered tests take so long to get results. Here we discuss some of the common reasons that tests take longer to perform and report than non-laboratorians may think it should.

Different tests take different amounts of time to perform. While this may seem obvious, the time it takes to get a test result can range from minutes to weeks or even months. Many hospital laboratories offer a test menu with hundreds of different analytes. If you've read Chap. 4, you'll now appreciate the wide variety of methods used to generate laboratory results. There are tests that take very little time at all, such as a urine dipstick test which measures nine parameters simultaneously (specific gravity, leukocyte esterase, nitrite, protein, glucose, ketone, bilirubin, blood, and urobilinogen). In principle, these single-use dipsticks specially treated with chemicals is dipped into a sample of clean-catch collected urine. Each of the chemical squares on the dipstick changes color and is then compared to a color-coded result chart on the dipstick container within minutes. The results can be recorded within minutes and entered (or sent digitally if there is a dipstick reading device) into the laboratory information systems (LIS), which then communicates the result to the hospital information systems (HIS or electronic medical records). In our laboratory, the turnaround time (the time you would expect a results) for a routine urine dipstick is 2 h. The simplicity of this test parallels the speed with which the lab is able to return test results. As you may have gathered from the preanalytical chapter (Chap. 3), most of the lapsed time is spent delivering the sample to the lab, accessioning it (processing the test order requisition, confirming the correct identification of the patient, etc.), and then moving into the cue for analysis. For most physicians, the clock begins to tick after they place the order, but a great many events need to happen before the laboratory technologist analyzes the sample.

In contrast to the urine dipstick test, there are tests that are much more complex, involve multiple steps, and are labor intensive. Because they are more complex, they take longer to perform, and therefore, take longer to get a result back to the ordering provider. For example, amino acid analysis by liquid chromatography/tandem mass spectroscopy (LC-MS/MS) involves extracting the amino acids from the plasma, urine, CSF, or even tissue extracts before being analyzed. To expand on the mass spectrometry methodology described in Chap. 4, a small amount of the sample is mixed with an acid solution to precipitate any proteins in the sample; a buffer is added to achieve a basic pH for a labeling reaction. Samples are labeled with a reagent for 30 min, dried, and mixed with the internal standards pre-labeled with the same reagent. This mixture is then injected into the LC-MS/MS for analysis; it is first separated by column chromatography where a solution gradient, wash, and equilibration take a total of 20 min. The results are then analyzed, compared to amino acid standards, identified, quantified, and finally reported. The 20 or more steps involved in this process, including precise pipetting mixing with internal standards accounts for the much longer time this test requires to be completed. In between the very quick urine dipstick and the much longer LC-MS/MS quantitative amino acid analysis, there are many tests that are intermediate in complexity, and therefore intermediate in their turnaround times (see Table 5.1).

Table 5.1 Turnaround times for common laboratory tests

	Routine	Stat	Analysis time
Arterial blood gas	15 min	15 min	2–3 min
Urine dipstick	2 h	1 h	3 min
Complete blood count	2 h	30 min	3–5 min
Routine chemistry panel	2 h	30 min	3–5 min
Anti-neutrophil cytoplasmic Ab (ANCA)	1–7 days	NA	2–3 h[a]
Urine/serum electrophoresis	2 days	NA	1–2 h[a]
Amino acids, quantitative, plasma	3–5 days	NA	3–12 h[a]
Hemoglobinopathy profile	3–5 days	NA	2–3 h[a]

[a]Require additional time to interpret results, usually once a day

Some tests have few samples submitted and therefore are batched. For most tests, controls are run at least once a day (many times more often) to ensure that the test itself is performing within the required limits. Sometimes control reagents are required every time samples are run. So when few patient samples are submitted at a time, they are batched; i.e., they are collected for certain periods of time (i.e., 2–3× per week) and run all at once. In this way, the control reagents and technologist time are used more efficiently. For example, liquid chromatography-tandem mass spectrometry (LC-MS/MS) is commonly used for monitoring therapeutic drugs, such as tacrolimus, and is batched because this is not a low volume test with an extraction step. In contrast, chemistry tests and complete blood counts continuously come into the laboratory and are run all the time and therefore are not batched. One can determine if a particular test is batched or not, by looking at the laboratory's posted turnaround times and test availability. For example, in Table 5.1, both amino acids and hemoglobinopathy profiles can be seen as batched because they are run only so many times per week. Similarly, urine and serum electrophoresis analysis is batched daily for this reason. If a test is only offered during daytime hours during the week, it may also be a batched test. If it is not clear which tests are batched or not, contact the laboratory directors to find out. You may also consider contacting the laboratory if the turnaround time is excessive as in some cases there is flexibility to coordinate analysis days with clinic visits.

Some tests are not done locally and they need to be sent out. Whereas hospital in-house menus typically include a few hundred tests, the use of reference laboratories (introduced in Chap. 2) allows for a menu of over a thousand tests. There are a number of large reference laboratories in the United States (see Table 5.2) that run a vast array of both common and esoteric tests. They perform tests on such a large scale that they are often able to provide low volume tests relatively inexpensively.

The first priority when determining if a test is sent to a reference laboratory is clinical need. For example, it makes no sense to send troponin to a reference laboratory for acute management of patients with acute coronary syndrome. If the clinical need is not immediate, then there is a cost analysis comparing in-house prices to those of the references laboratories (see Laboratory Budgets in Chap. 1). The income that can be generated by offering these tests is limited, as the "markup" or

Table 5.2 Reference laboratories in the United States

ARUP Laboratories (http://www.aruplab.com/)
Laboratory Corporation of American (LabCorp) (www.labcorp.com)
Mayo Medical Laboratories (http://www.mayomedicallaboratories.com/)
Quest Diagnostics (www.questdiagnostics.com)

While not an exhaustive list, these websites can give you an idea of the vast array of testing that is available

profit margin is generally small; at public hospitals there are a number of patients/insurance companies that default on payment. So when a laboratory is only running a few tests a week, it rarely is worthwhile from a cost standpoint to bring these tests "in-house." Examples of tests that have met the cost threshold for in-house testing are serum free light chains and vitamin D. When it became standard to run serum free light chains on every multiple myeloma patient and vitamin D levels with every healthy physical, the equation changed dramatically. With a greater utilization of these tests, the relative cost of the equipment, control reagents, and medical technologist time became much smaller than the reference test costs. At this point the tests are brought in house.

If you are ordering a test that is not commonly requested, it is likely that your laboratory will send it out for referral testing. There is generally a "Referral Testing" area in hospital laboratories that complete the paperwork, properly prepare samples for overnight shipment, and then ship the samples to the reference laboratories. In some cases, the laboratories are physically very close to the hospitals, so they routinely pick up samples multiple times a day, resulting in a quick transport time. Since sending out tests is a labor-intensive process requiring courier transport, they commonly take days to weeks to complete. Fortunately, many of the large reference laboratories have laboratory information systems directly connected to the referring hospital, allowing the results to be transmitted back immediately upon completion, to then go into the medical record. However, not all laboratories have this infrastructure in place, resulting in a bit longer turnaround time. It is important to note that rare tests are generally much more expensive than in-house tests. Referral tests are frequently a target of test utilization initiatives as they can quickly eat up a laboratory budget; compare pennies for creatinine to thousands of dollars for a molecular test. Before ordering an uncommon test, it is worth considering whether it is needed and how you might interpret the results if they come back normal, abnormal, or equivocal. If you don't know the answer, find out *before* you order it. Likewise, if you feel that a test is needed locally, you'll be able to more effectively communicate with the laboratory if you have an understanding of how the in-house test menu is selected.

Another reason why samples are sent out is that a given test may only be available in one laboratory in the country. For example, serological testing for Toxoplasmosis is only performed by the "Toxoplasma Serology of the Palo Alto Medical Foundation Research Institute" (www.pamf.org). This laboratory has been the only laboratory that offers an array of testing options for Toxoplasma (IgG, IgM,

IgA, IgE, Differential agglutination (AC/HS), etc.). When only one laboratory offers an esoteric test and it cannot be brought into your hospital, then sending out the samples for analysis is the only choice, and it will inevitably take longer because of the send out process itself. Note that even the largest of reference laboratories are likely to batch rare tests for the same cost-driven reasons as hospital laboratories. Reference laboratories also send some tests to other reference laboratories, such that a specimen may be transported twice before reaching its destination testing laboratory!

Some test methods involve a multi-day step preventing them from ever being done quickly. Some tests involve steps that take a long time, which makes the results take longer to be reported. Good examples of this are radioimmunoassays and cold agglutinin tests. While radioimmunoassays (RIA) are not used as commonly as they were in the past, they are still used for some endocrine tests (e.g., renin activity and adrenocorticotropin). While these are very sensitive tests (i.e., can detect very low hormone concentrations reliably), which is why they are still preferred, they often have incubation periods that last 24, 48, or even 72 h. Another example is the identification of cold agglutinins. To identify cold agglutinins, samples are precipitated overnight, generally being cooled to 30 °C or lower. The precipitate is then analyzed by immunofixation electrophoresis, which itself takes several hours and requires visual interpretation. With these long steps, the test takes longer, and therefore so the turnaround time.

Quick tips to the "Why is my test taking so long" question. With the vast array of testing modalities available both within the hospital and through referral testing, how are you going to learn the details of each of the laboratory tests you order? There are a couple of tricks that we think might help. First, realize that the laboratory is on your side and is a resource that can help you—that's why we're here. As a result, most laboratories have spent considerable time, effort, and money developing web resources that can answer many of the questions you might have about tests. For example, in our hospital, we have a continuously updated test menu (http://labs.unchealthcare.org/labstestinfo). On these pages, the type of tube needed to collect the appropriate sample is given, as well as the routine turnaround time, the availability of STAT testing, and details indicating if its batched or not. For example, for serum protein electrophoresis, blood needs to be collected in a 3 mL serum separator tube (gold top) and is reported by 4 p.m. Forty-eight hours after receipt (i.e., it is batch tested). The laboratory turnaround times are benchmarked for accreditation purposes, so >95 % or more of samples (on average) will meet these times. Lastly, websites list tests commonly sent out to referral labs. For example, if you look up Cold Agglutinins in our hospital, you will see that they are collected in 5 mL red top tubes and sent to a reference laboratory (Mayo); the estimated turnaround time is 4–6 days. The laboratory is there to support you, so if you have any questions about why your test results are not available, always feel free to contact lab staff directly. To find out more information use your lab's website or laboratory manual as the answers to your questions may already be at hand.

Why Is the Laboratory Calling Me?

Patient care is the goal of the laboratory even though we may never see the patient. As we are focusing on specimen testing, we are cognizant of the fact that critical, often life and death decisions are being made based on the accuracy of lab work. So anything that the lab can do to improve patient care is a top priority. Having said that, it may come as a surprise that the lab sometimes needs help from the ordering providers to optimize the testing process. This is the primary reason that the laboratory contacts physicians and other medical providers. They are not second-guessing the front-line team; they are only trying to provide the best care for the patients. Here we discuss some common reasons that the laboratory contacts physicians and other medical providers about their patient's lab tests.

Important information needed to run and/or interpret the test is missing. A common example of this in the recent past was the need for a patient to sign a consent form to have an HIV test run. Without the proper consent, it was not legal to perform and report the results of the test. Other tests require additional information to interpret them. For example, running the AFP maternal quad screen requires a maternal serum request form, gestational age, race, diabetes status, weight, and family history. Without all these pieces of information, the results of the test are uninterpretable. So if this information is missing, the laboratory will be calling to get all the necessary information.

The wrong specimen was submitted for the test ordered. Submitting the wrong specimen for the test that was ordered is another reason that the laboratory would contact you. Since the test was important enough to order, the laboratory wants to fulfill its commitment to the patient by completing the order properly. So if you (or your support staff) submitted an order for ionized calcium in a serum separator tube (red top) instead of the required 1 mL green top tube, you will be hearing from the laboratory. Since tubes lacking the heparin in the green top tubes will not give accurate test results, it is critical that the correct sample is ordered.

Another potential issue is that there may be a better specimen type to help make a diagnosis. For example, acute pulmonary histoplasmosis is a fungal disease that results in nonspecific respiratory symptoms including a cough resulting from exposure to bird or bat droppings. It can disseminate and affect multiple organs and can be fatal, particularly in immunocompromised patients. While there are tests for serum antibodies for *histoplasma*, and blood cultures can grow the organism in 6 weeks, the much more specific and sensitive test is urine antigen testing. It is not uncommon for clinicians to order tests for *histoplasma* not realizing that the best and fastest test is the urine antigen test. Therefore, the laboratory may call to make this point, requesting that the appropriate specimen (urine) and test (urine antigen testing) be ordered. Again, it significantly improves the care of the patient and the contact is to help you in managing your patient. Nobody expects the end test users to memorize the multitude of sample requirements for each test, that's why the lab is there to help. But, the lab can't do the work properly without the right material collected and ordered the right way.

The test ordered is not appropriate or does not include the right tests to allow interpretation. There are times when clinicians suspect a rare cause of disease, but are not sure what the best testing strategy is. They may search the laboratory test menu or search the Internet for tests that are available to assist with their decision-making process. While there may be many possible tests to diagnose a disease, there are many that have little to no value, particularly when multi-tiered testing is recommended.

One example that we have seen in our institution is with the diagnosis of porphyrias. Porphyrias are a group of eight rare inherited metabolic disorders of the heme biosynthesis pathway. In acute cases, neurological dysfunction can affect the autonomic, peripheral, or central nervous system (Cappellini, Brancaleoni, Graziadei, Tavazzi, & Di Pierro, 2010; Sassa, 2006). Autonomic neuropathies are common and manifest as gastrointestinal symptoms (abdominal pain, nausea, vomiting, constipation, diarrhea, tachycardia, or hypertension). Peripheral sensory-motor neuropathies include pain and numbness in the extremities, muscle weakness, and acute or chronic paresthesia. Central nervous system involvement includes convulsions, confusion, anxiety, depression, and insomnia. All of these symptoms, such as abdominal pain, are fairly common, so porphyrias get on the differential diagnosis of diseases fairly commonly.

The diagnosis of the different porphyrias potentially includes up to seven different tests (e.g., whole blood porphobilinogen (PBG), feces porphyrins, urine organic acids screen). And many times physicians order these tests first, disregarding the difficulty of some of the tests (feces). Fortunately, testing algorithms have been developed to most accurately and efficiently screen for all eight porphyrias, starting with quantitative porphyrins from random urine and a 24-h urine aminolevulinic acid (ALA) in patients with the appropriate symptoms of Porphyria (see http://www.mayomedicallaboratories.com/media/articles/algorithms/acuporphyria.pdf).

A common reason that the laboratory, specifically pathology residents and clinical fellows, then call the ordering physicians is to help guide them through this process. A common scenario in the case of porphyrias is that a physician has a patient with intermittent pain, so they order the PBG when the patient is not having the intermittent pain. There are a host of reasons why this might miss the diagnosis, whereas the recommended screening of quantitative porphyrins from random urine and a 24-h urine aminolevulinic acid (ALA) during an acute crisis would be the most efficient way to make the diagnosis, or exclude it. Since these are generally send-out tests because of their low volume and cost, it might take a couple of days to get the result. If the wrong test is ordered, the process of making the diagnosis is then delayed that much longer if additional testing is needed. Again, the overarching goal of the laboratory is to help the clinicians make the most efficient and accurate diagnosis, for the purpose of helping the patient.

Critical values. Most laboratories have policies requiring that staff notify the ordering physician of critical values. Critical values are defined as results that represent a pathophysiological state so far out of normal as to be life-threatening by itself if something is not done immediately to correct the issue. Critical values do not *necessarily* correspond to a normal reference range, therapeutic ranges, or even toxic ranges and not all tests have critical values. Laboratory websites or reference manuals are a good resource for the critical values of individual tests. For example, WBC counts have critical values of <0.5 and >50 × 10^9 cells/L. In both clinical scenarios, there is a high probability of severe clinical complications including immunocompromisation (<0.5) or acute leukemia (>50) requiring immediate intervention.

Even though large hospital and reference labs serve a diversity of different clients, there may be only one defined “Critical” result for all patients for a given test. Consider glucose critical results. In an ER population, glucose may be commonly elevated to critical levels in patients experiencing high stress or trauma (see Preanalytical Variables Chap. 3). In this situation, ER staff might find repeated calls with critical results uninformative or even distracting. However, if the critical threshold was adjusted to avoid calling too many critical results to the ER, then ambulatory patients with critical glucose results could have recognition of their life-threatening condition delayed. Establishing critical result flags is a continuous balancing act between signal and noise. Technology can help create location-specific critical results, but it isn’t always as flexible or available as needed for patient care.

Because the laboratory is subject to regulation, there are typically policy statements requiring that the provider is notified when critical limits of specified tests are exceeded and/or critical results are obtained. The laboratory staff must contact the ordering physician within 60 min following the release of the critical result. To verify that accuracy of patient information communicated by phone, the physician or designees is required to read back the patient name, unique patient medical record number, and the critical test result. The laboratory then documents this interaction including the patient information, the critical result, the person they talked with, the time, and confirmation that an accurate read-back of the information occurred. In line with many of the other functions described in previous sections, the laboratory may call you with critical values designed to help assist in critical decision-making for your patient.

The laboratory was asked to call with the results. Sometimes the reason the laboratory calls is as simple as they were asked to do so. This may happen when the laboratory test is critical to the treatment of the patient. For example, toxicology screening may help assist the treating team as to the cause of stupor, therefore allowing them to counter-act it pharmacologically as soon as possible. We have also been asked to call ordering teams for the results from a CSF immunofixation test for the identification of IgG oligoclonal bands so that the diagnosis of multiple sclerosis could be made as soon as possible and the appropriate treatment started. When there are difficulties with tests, such as interfering substances or the need to large dilutions, we are also asked to call the ordering team to give the results. When an influential member of a specialty makes a request of the laboratory, the entire clinical service may be subject to its delivery.

The strange thing about working in teams taking care of patients. While we’ve hinted at the fact that communicating between multiple large teams to take care of patients is challenging, it is one of the reasons you are getting phone calls from the laboratory. So even though you might not know why the laboratory is calling, it is most likely there is good reason that involves the optimal care of your patient.

Why Haven't I Received My Results Already?

So you've become savvy and understand the rationale for the different times it takes for different tests. You've looked up the turnaround times and realize that double that amount of time has passed than would be expected for a particular test that is being run in your laboratory. Here we discuss possible reasons for additional delays and develop ideas on how you can communicate with the laboratory to solve these problems together.

Reasons specimens are rejected or not analyzed. If the integrity of the specimen has been compromised, tests are not performed. Laboratories are adamant that no result is better than the wrong result. There are a number of common reasons for "specimen rejection" as outlined in Table 5.3.

Solution(s): Re-collect the sample and submit it in the appropriate tube considering the appropriate requirements of the sample. The laboratorians/manufacturers that have developed each test have extensively identified the conditions in which the tests give accurate results. Regulatory and validation processes have specific guidelines on the performance of each test, so the sample requirements must be met at all times as part of overall quality. If the laboratory accepted these specimens, there is a high probability that the results are inaccurate, negating any value the test may have had. This is a very difficult situation when rare samples are collected or collections are from babies with a limited amount of blood that can be drawn. However, it is in the best interest of the patient for the laboratory report accurate results, not just any results.

Minimum sample volumes are needed to run tests. One of the reasons samples are not run is there is an insufficient amount of the sample (flagged as quantity not sufficient or "QNS") needed to perform the test. The laboratory did not necessarily spill your sample resulting in the QNS! There are multiple reasons why you or your support staff may not have provided an adequate sample amount, even though it may have seemed generous.

First off, if testing requires serum or plasma, then roughly half of the submitted volume represents red cells, which are not analyzed; as a rule of thumb, take the total sample volume and subtract the hematocrit. Secondly, many laboratory tests are processed almost completely by machines. If there is too little sample, the probe will suck up air, and will give an error so that the test won't be run/resulted. When samples are extremely short, the probe can enter the gel (in serum separator tubes), resulting in damage to the probe, and long delays in testing due to the need to repair the instrument itself delaying reporting of other results.

Table 5.3 Common criteria for specimen rejection

Specimen collected in the wrong tube or container
Incorrect information (not dated, name misspelled, unsigned)
Specimen inappropriately handled (inappropriate temperature, timing, or storage requirements)
Quantity not sufficient (QNS)
Lipemic or grossly hemolyzed specimens (for certain tests that these are known to be interferences)

Automated systems also have a "dead volume" for each sample. The dead volume is that which fills up the tubing that delivers the sample to the testing area (generally a small cup). In addition to the physical requirements of the probes in automated samples, the amount of sample needed is optimized for specific volumes based on their ability to detect small quantities of analyte. Most tests are designed optimally with the smallest amount of analyte needed for analysis. So if you do not have an adequate amount of analyte, the test will not give you an accurate measurement, negating the value of running the test in the first place.

Sample quality issues (*interferences and contamination*). In Table 5.3, we outlined some of the reasons that specimens are rejected. The last criterion for specimen rejection involves known interferences, including hemolysis, icterus, and turbidity/lipemia.

(a) *Hemolysis*. Hemolysis can occur during phlebotomy or may be the result of a pathophysiologic process, such as autoimmune hemolytic anemia or secondary to a transfusion reaction. Hemolysis affects different tests on different instruments in unpredictable ways. *It cannot be assumed that that because of hemolysis, laboratory results are increased or decreased.* There are generally cutoff values for specific tests for the amount of hemolysis that may be tolerated without affecting values. However, most significant hemolysis prevents the accurate measurement of tests, completely negating the value of analyzing them. This is why visible hemolysis is used as a reason for rejecting specimens. For example, the Clinical and Laboratory Standards Institute guidelines for prothrombin (PT) and activated partial thromboplastin time (aPTT) state, "Samples with visible hemolysis should not be used because of possible clotting, factor activation and interference with endpoint measurement" (CLSI, 1998).

An extreme example of how hemoglobin can affect most laboratories was seen during the development of blood substitutes in the past decade. These blood substitutes have been based on using purified hemoglobin. Since pure hemoglobin separated from cells causes renal toxicity, different ways of modifying it have been developed, including chemical cross-linking, polymerization, or encapsulation. The use of cross-linked hemoglobin blood substitutes has been tested in clinical trials; samples from these patients, they appeared excessively hemolyzed and significantly affected results, in some cases yielding no result at all.

While hemoglobin-based oxygen carrier (HBOC) concentrations up to 50 g/L did not interfere with Na+, K+, Cl−, urea, total CO_2, Mg, creatinine, and glucose, tests for total protein, albumin, LDH, AST, ALT, GGT, amylase, lipase, and cholesterol were significantly affected preventing their accurate determination (Ma et al., 1997). CK-MB, CK, GGT, Mg, and uric acid were significantly affected at even low concentrations of HBOC in other studies (Wolthuis et al., 1999). This was quite surprising to both the clinical investigators that wanted to follow the effects of HBOC on their patient's laboratory values, and certainly a surprise to the laboratory directors that were challenged with enormous amounts of hemolysis. It remains apparent that blood substitute manufacturers seldom consult/consider the laboratory during development, so this remains a potential future challenge.

(b) *Icterus and Lipemia.* Icterus is caused by increased levels of circulating bilirubin, which can be secondary to in vivo hemolysis or decreased ability to conjugate and clear bilirubin via the liver. Common causes of lipemia include diabetes mellitus, alcohol abuse, chronic renal failure, pancreatitis, cirrhosis, systemic lupus erythematosis, and medications including estrogen and steroids (Ji & Meng, 2011).

While icterus and lipemic interferences generally do not affect as many tests as hemolysis, they do interference with many tests. For example, one recent study found that CO_2, ALT, Albumin Ca2+, CK, creatinine, GGT, HDL, total protein, and uric acid were all affected by interferences (Ji & Meng, 2011). Icterus interfered with more immunoassays than hemolysis or lipemia (hCG, IgG, and Free T3) and more drug assays (acetaminophen, gentamicin, phenobarbital, theophylline, tobramycin, and vancomycin) (Ji & Meng, 2011). It is because of these widespread interferences that samples with increased icterus and lipemia are not reported. For options with patients that have medical causes of these clinical laboratory interferences, it is recommended that you contact the laboratory directors for guidance in proceeding to get accurate measurements of needed analytes in these patients.

Another important point regarding interferences is that they are often method specific. This means that when instrumentation changes, so will the susceptibility to a given interference and therefore the rejection/acceptance criteria.

Interpretation of results with preanalytical issues (potassium as an example). Clinicians should be aware that there are a number of reasons for falsely increased potassium (Baer, Ernst, Willeford, & Gambino, 2006).

(a) *Hemolysis.* Red cells contain large amounts of potassium (23 times the amount of K+ in circulation), so they release potassium when damaged (hemolyzed). Hemolysis can result from excessive suction applied to a syringe plunger, forcibly ejecting the blood sample from a syringe into an evacuated tube, and drawing the blood through a small-bore needle or catheter (Van Steirteghem & Young, 1977)
(b) *Potassium released from platelets and red or white cells.* Red cells, white cells, and platelets all contain relatively high amounts of potassium. Thus, there are numerous scenarios, where they can release potassium resulting in falsely elevated levels. For example, potassium may be falsely increased if serum is allowed to clot >2 h, if there are delays in processing, or if centrifugation exceeds recommended time or force (~$1200 \times g$ for 10 min) (Baer et al., 2006). This is aggravated in cases of leukocytosis or thrombocytosis.
(c) *Specimen contamination.* Contamination can cause falsely elevated results in two main ways: (1) Potassium can be introduced directly into the sample or (2). Compounds contaminating in the sample can cross-react with the ion-selective electrode and be interpreted as potassium (Baer et al., 2006). Potassium can be falsely elevated if the collection tubes are filled without regard to their order.

If blood collection includes a potassium-EDTA (K2EDTA) tube, it can carry over into tubes later in the draw sequence. A specific tube draw order has been developed in order to prevent this from happening (see Draw Order in Chap. 4). Povidone-iodine disinfectant used to disinfect skin can also cause elevations in potassium if they contaminate the draw (Van Steirteghem & Young, 1977). Benzalkonium-heparin bonded catheters are used in critical care areas to prevent thrombi from forming; however, the surfactant properties of the benzalkonium chloride interfere with the potassium and sodium specific electrodes (Gaylord, Pittman, Bartness, Tuinman, & Lorch, 1991; Koch & Cook, 1990). Since this coating is eluted early after its use, contamination can be avoided by flushing the catheter with 10 mL of blood before drawing the potassium sample. Likewise drawing from a line from which the patient is receiving IV potassium will cause falsely elevated values (see Chap. 4).

(d) *Fist clenching*. Fist clenching or "pumping" before or during venipunctures, as has been taught to generations of medical students and phlebotomists, adversely affects the potassium by falsely elevating it; the mechanism is proposed to be local release of muscle potassium in the forearm (Don, Sebastian, Cheitlin, Christiansen, & Schambelan, 1990).

(e) *H+/K+ ion exchange*. Hyperventilation and crying can either increase or decrease potassium levels depending on the time of the episode. Hyperventilation for 3–6 min can cause a rapid uncompensated alkalosis and rapid shift of potassium to the serum (potassium moves in the opposite direction of hydrogen ions) resulting in hyperkalemia. After 30 min or more, the body compensates for the low H+ and potassium shifts back into the extracellular space, which may cause an apparent temporary hypokalemia (Baer et al., 2006).

Interpreting Laboratory Results

There are myriad medical textbooks, pocket guides, handbooks, and websites dedicated to interpreting laboratory tests. However, few of these provide the laboratory perspective, which can augment a physician's integration of test results into patient care. So rather than delving further into the pathophysiology of disease, let's start with the practical information provided by the laboratory in result reports. Depending on the institution and the laboratory service, results reports range from a digital output with rows and columns of results generated from a relational database to a faxed paper print out. While these are vastly different on the surface, laboratory regulations (see Chap. 1) require that certain information be found with each type of report. At minimum, these include patient identifying information, the ordering physician/location, reference intervals, and the name of the laboratory that did the testing. While this may get lost in the flood of information flowing to the physician, it is essential to appreciate what all these mean and how they can be used to the advantage of the care provider.

Reference Intervals

Reference intervals (described in Chap. 4) serve as a comparison point for the patient results. However, is it the right comparator for your patient? As you may recall from Chap. 4, reference intervals are usually generated by studying healthy individuals who are willing/available to participate in a study. In a hospital lab, this may include a predominance of young or old women, reflecting the most abundant type of employee (this varies between labs). If you consider that reference intervals are generated from a population that may not represent your patient, then one needs to reconsider what constitutes an abnormal result.

Some examples of this are at the extremes of age. There is woefully little information on what constitutes "normal" for aging adults in particular. Is there a 90-something year old out there without some type of comorbidity? If so, can you get a few dozen different 90-year-olds to do your reference interval study? What about 10-day-olds? In the case of age-specific reference intervals, there simply may not be enough information for some tests in a particular age range. Consider that the amount of data needed to generate reference intervals for each age group increases with each age/sex partition; the size of the normal value study essentially doubles for each partition added. If 120 patients are needed to get a hemoglobin reference interval for the general population, then 240 will be needed to generate the interval for males and females.

It is noteworthy that some gaps are in the process of being filled, as there are concerted efforts to generate robust reference intervals for pediatric patients (http://www.aacc.org/resourcecenters/resource_topics/pediatric_reference_range/; http://www.caliperdatabase.com/caliperdatabase/controller; http://www.nationalchildrensstudy.gov/Pages default.aspx). Even still, reference intervals are somewhat brittle when it comes to changes with age as illustrated in Fig. 5.1. Geriatric reference intervals remain works-in-progress as do reference intervals for different ethnicities, and patients with comorbidities. One question that arises from Fig. 5.1 is: "When a patient turns 10 does something dramatic happen overnight to their ALP levels?" Many computer systems used to present laboratory information are highly limited, and can't display continuous reference intervals (smooth transitions between ages as opposed to the jarring changes shown in the figure). While there are ongoing efforts to improve this situation, reference intervals ultimately serve as a guideline. Clinicians need to rely on other information with the clinical context to determine how the results fit into their differential diagnosis.

In some instances there may not be any reference interval provided whatsoever. This may be the case with protein or electrolyte measurement in random urine samples. Because of the huge variability in urine concentration, the range of possible results is so wide and sample-dependent that it ceases to be informative. Of course in other cases, reference intervals from healthy individuals are simply inappropriate; for example, how much acetaminophen should someone usually have on board? For some analgesics, including acetaminophen, there are simply cutoffs for positivity, where the clinician needs to interpret the concentration in concert with the clinical presentation and the timing of ingestion (e.g., Rumack-Matthew nomograms). For therapeutic drugs there are target ranges, which in the case of immunosuppressants

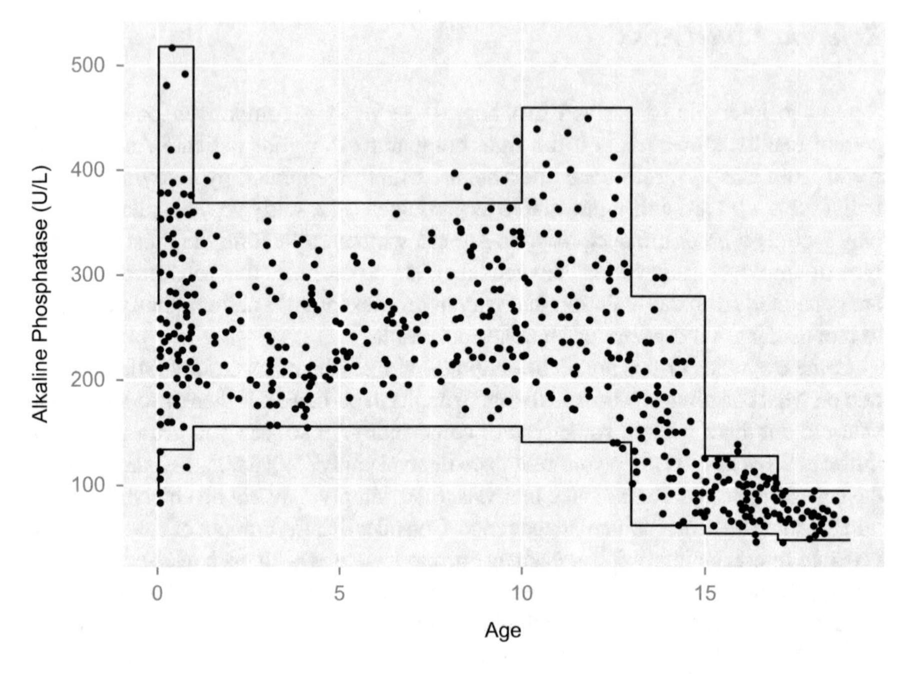

Fig. 5.1 Alkaline phosphatase age-based reference intervals. Values of alkaline phosphatase in children enrolled in the CALIPER study (http://www.caliperdatabase.com/caliperdatabase/controller). *Upper* and *lower lines* indicate the reference intervals determined from the data

are transplant-type dependent and require careful collection timing to interpret. For other tests, such as cholesterol and HbA1c, there are nationally or internationally defined targets derived from evidence-based medicine, such that a "normal" range is not informative.

As with most things communication is beneficial. Laboratories will inform physicians of changes to reference intervals and physicians should tell laboratories if the intervals appear incorrect. For small community hospitals, it may not be possible/practical to get enough people to validate age-partitioned intervals, so the manufacturer's reference intervals are frequently adopted. With local population differences, physician input to the laboratory is useful when results are unexpectedly abnormal on a recurring basis. Laboratories are also required to reevaluate reference intervals on an ongoing basis. If instrumentation or the population changes, then references intervals may also need to change.

Result Flags

Additional information found in laboratory reports includes result flags. The main types of result flags are "Low," "High," and "Critical." As described earlier in this chapter, "Critical" results are a predefined set of cutoffs agreed upon by the physician leadership at the hospital and the laboratory. These flags will trigger a phone call and have special handling considerations. Less urgent than critical flags are "Low" and "High" result flags, which simply indicate that results are below or above the reference interval. These results need to be interpreted in the context of the physical signs and symptoms using clinical judgment. Again, consider that reference intervals are defined based on 5 % of healthy individuals having an "abnormal" test result. Therefore, if a patient has 20+ tests, then it is probable that at least one will be abnormal in the absence of disease.

Biological Variation and Interpretation of Laboratory Results

Going beyond a single result, it is worth considering change in laboratory results over time. This section will expand on the concept of biological variation (discussed in Chap. 3) from a post-analytical standpoint. As described in Chap. 4, biological variation refers to the amount an analyte varies within an individual or group over time. When interpreting laboratory data one must consider what is "normal" for a given patient. Some analytes are highly individual, whereas others tend to be similar between individuals. An analogy is an individual's body mass index, which doesn't tend to change much within an individual in a short time, but is quite variable between individuals. It is useful to consider this concept with common laboratory values (Fig. 5.2). The figure shows computer-simulated examples of 1000 results from individuals and groups of individuals based on published biological variation (Callum, 2001). Hemoglobin and creatinine are considered highly individualized laboratory results, where the difference within an individual is small in comparison to the difference between individuals. In comparison, iron and potassium display similar distributions within and between individuals. For highly individualized analytes, when a result is within the reference interval, it may not necessarily be "normal" for a given patient. Conversely, when a result from a highly individualized analyte is outside the reference interval, it is highly likely to be abnormal for a patient.

An extension of this concept, which aids in the interpretation of laboratory results, is the reference change value (RCV). The RCV helps answer questions such as: Is a change in potassium from 3.5 to 3.3 mmol/L in 24 h significant? Does a decrease in hemoglobin of 20 % over 2 days represent noise or real change? To answer these, one needs to consider several factors beginning with the sample. If a patient is getting DW5 + KCl (5 % dextrose with 0.9 % saline and 20 mEq potassium chloride) and the sample is contaminated with some of this fluid, then the results may be altogether uninterpretable. If preanalytical considerations are satisfied, then the concept of reference changes values becomes useful. Reference change values refer to the amount (%) by which serial results must differ to represent clinically significant change. Reference change values are calculated from analytical imprecision[1] and biological variability data. Described in Chap. 2, biological

[1] *Analytical imprecision refers to the amount of "wobble" there is in successive measurements of the same sample; for instance, if potassium were measured ten times in a row in a sample with a concentration of 3.5 mmol/L, measurements would range from ~3.4 to 3.6 mmol/L; see Chap. 4.*

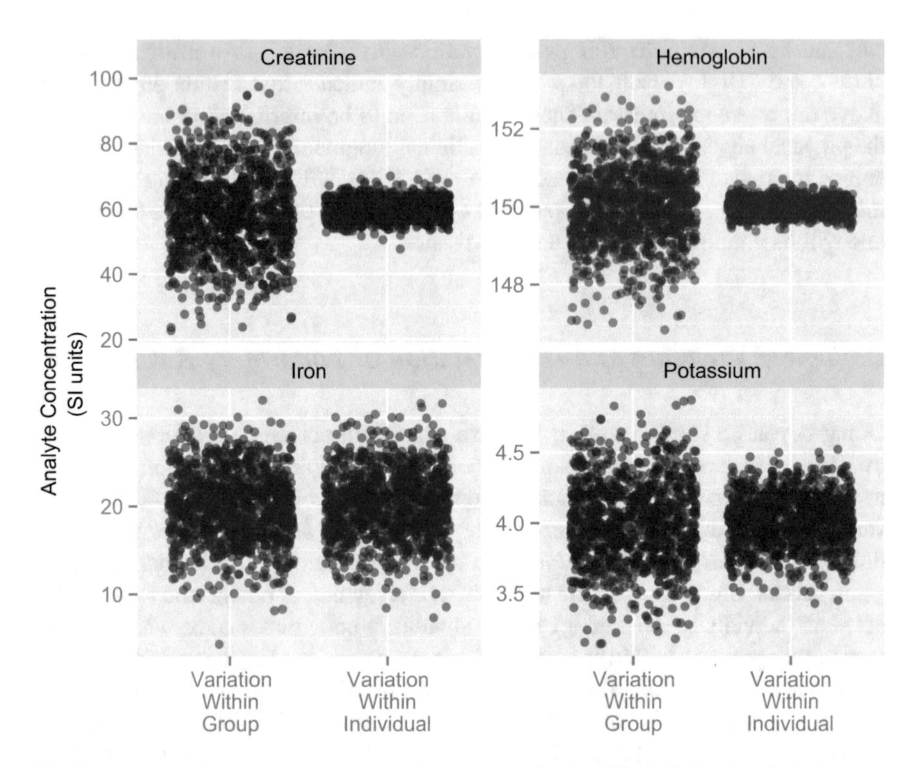

Fig. 5.2 Biologically variation simulation. Data are simulated within individual and within group variation for common analytes. 1000 points were randomly sampled from a normal distribution based on published biological variation data (Callum, 2001)

variability is the amount an analyte changes in an individual or between individuals over time. The reference change value can be calculated using this simplified equation:

$$\mathrm{RCV} = 2^{\frac{1}{2}} \times Z \times \sqrt{\mathrm{CV_w}^2 + \mathrm{CV_a}^2}$$

The value $2^{1/2}$ is a constant derived from the fact that we're comparing the difference between two values. The Z is usually 1.96 for a 95 % confidence level, CV_w is the within individual biological coefficient of variation, and CV= is the analytical coefficient of variation. With this simple equation, it can be calculated whether the change in the result real or noise. Continuing with the example of potassium, which has a CV_w of 4.8 % and an analytical variation (CV_a) of 2 %, the reference change value is 14.2 %. Therefore any change in the potassium concentration of >~15 % or more represents real physiological change. Thus, to answer the question above (Is a change in potassium from 3.5 to 3.3 mmol/L significant?), an increase or decrease of 0.2 mmol/L in a patient's potassium is <10 % and is therefore more likely noise than true physiological change.

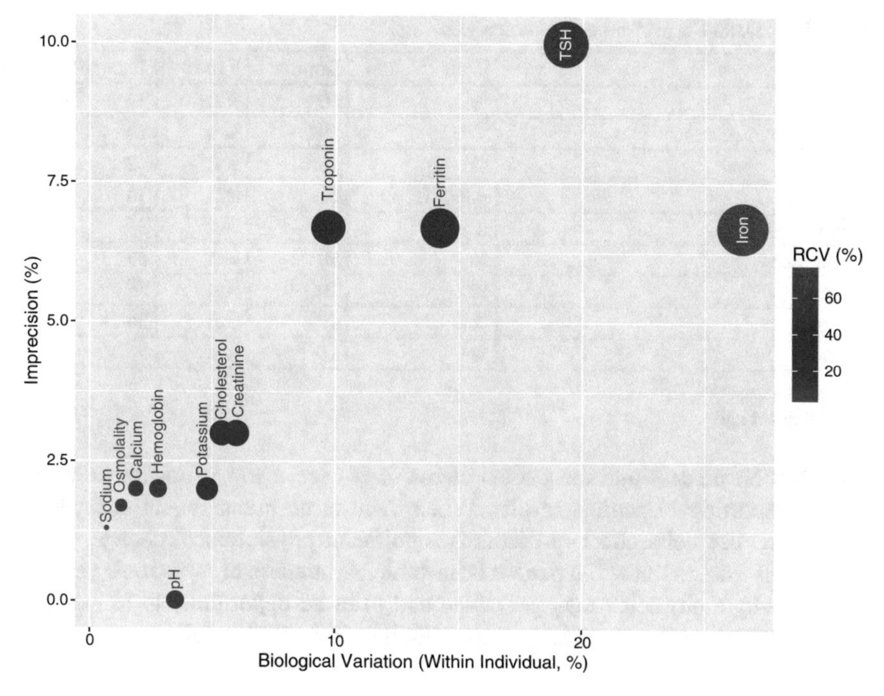

Fig. 5.3 Effect of biological variation and analytical imprecision on reference change values for common analytes

From the reference change value equation, it can also be inferred that greater analytical imprecision contributes to larger reference change values (illustrated in Fig. 5.3). While electrochemical methods, such as potassium, are highly precise, those of immunoassays for hormones tend to be more variable (CVs may be >10 % for some immunoassays). While this additional information is not likely to appear on a laboratory report any time soon, clinical laboratory directors will have analytical performance data, biological variation, and reference change values at hand, and may help interpret results.

While on the topic of lab value change, there is an additional result flag that is usually not provided to the physician called a "delta check." A delta check is an internal (lab only) flag where results for a given analyte have changed drastically over a short time. These are predefined rules that are designed to alert the laboratory staff to a problem with the sample (e.g., mislabeled or contaminated with IV fluid). These flags may prompt a phone call from the laboratory if they do not match the previous laboratory results (or other results on the same sample). An example of a delta check for potassium would be a change of ±1.5 mmol/L occurring in <24 h. In a stable patient, this amount of change is physiologically unlikely, so that results are flagged to ensure the sample did not come from another patient or as a result of contamination. Clearly there are pathophysiological events that can trigger this flag in the absence of a preanalytical error, but because the lab is blind

Table 5.4 "Delta Checks" for common analytes

Name	Units	Limit	Type	Time (h)
Albumin	%	50	%Δ[a]	72
Alkaline phosphatase	%	50	%Δ	72
ALT	%	100	%Δ	72
Calcium	mmol/L	0.3	|Δ|[b]	36
Chloride	%	10	%Δ	36
Creatinine	%	50	%Δ	36
Glucose	%	100	%Δ	36
Potassium	%	40	%Δ	36
Sodium	%	10	%Δ	36

[a]Percent change
[b]Absolute change

to collection process and the patient status, it is used a sort of insurance policy against reporting inaccurate results. When there is no preanalytical error, some institutions use delta checks to identify significant physiological change, where they might want to know if a patient is unstable. Awareness of these tools can help identify why a physician may get called and offer an opportunity to refine these flags to catch events that critical or low/high flags miss (Table 5.4).

Who Do I Ask About My Results?

With consideration of the laboratory structure described in Chap. 1, it may now seem obvious that different people in laboratory will be able to solve different problems. While calling anyone in the lab may get you to an answer, it may take a bit of time to navigate your way through to the person who can best answer a question. If the problem can be narrowed down in advance, depending on the system, you could directly contact the appropriate person. For example, if you simply want to know when a result will be ready, then you want to speak with someone in the area responsible for generating the results. If you want to see if the sample arrived, then the "front end" staff (also known as specimen processing, technicians, accessioning, clerks, or order entry staff) may be a good source of information. The "front end" staff are also usually a good source of information for the type of tube and volume of blood needed for a given test; they are the ones who receive and process the tubes and samples so they are very familiar with what tube type and sample-handling is needed for what test. When it comes to interpretative post-analytical questions or more obscure tests, it is best to contact the laboratory director or discipline specialist in charge a given testing area (i.e., a Clinical Chemist for Chemistry questions, Microbiologist for an infectious disease question, and Hematopathologist to interpret a blood smear). The person responsible for that laboratory testing site (Lab Directors and Specialists) will be able to answer questions about the source of reference intervals and result interpretation in clinical context. This is the reason the testing laboratory and contact information is provided (and required by regulatory bodies) on laboratory reports.

Laboratory directors have access to all the information in the laboratory including electronic medical records and can enlist the help of bench-staff where needed to answer questions analytically. Laboratory directors are specially trained and board certified to help physicians interpret results, and have a vast library of documents, books, papers, and technical manuals at their disposal. They are also usually well connected with other laboratory professionals who may be able to help in particularly challenging cases or when tests need to be referred out to another laboratory.

Likelihood Ratios and Test Interpretation

While considering test interpretation, it is worth taking a minute to consider what the results might mean before the test is ordered. Pretest probability refers to the expected presence of a particular condition before testing occurs. Consider an example where a patient presents with a sore throat and a RapidStrep test may be ordered or not. The pretest probability for an adult to have group A beta-hemolytic streptococcal (GABHS) pharyngitis (strep throat) is 10 % (Ebell, 2003). Using a RapidStrep with an assumed sensitivity and specificity of 78 %, the posttest probability is calculated as follows:

$$\text{Likelihood Ratio of a positive test}\left(\text{LR}+\right)=\text{Sensitivity}/\left(1-\text{Specificity}\right)$$
$$=0.78/\left(1-0.78\right)=3.54$$

$$\text{Post-test odds}=\text{pre}\quad\text{test odds}\times\text{LR}=10/100\times3.54=0.354$$

$$\text{Post-test probability}=\text{post test odds}/\left(\text{post test odds}+1\right)$$
$$=0.354/\left(0.354+1\right)=26\%$$

In this case, the likelihood ratio of a positive test is 3.54, which is readily calculated from the sensitivity and specificity. The posttest odds of a positive test are 0.354. This is more easily interpreted when converted to probability, which is 26.1 %. Thus, a positive test has only increased the probability of disease from 10 to 26 %. When the pretest probabilities are very low (<10 %) or very high (>90 %), tests have relatively little impact on the overall probability of a condition, such that there may be little value in ordering the test. If the performance of the test is poor, which may be the case with point-of-care tests in the hands of non-laboratory staff (Fox, Cohen, Marcon, Cotton, & Bonsu, 2006), then the diagnostic utility of the test decreases. Continuing with the RapidStrep example, in the hands of non-laboratory staff this test had a sensitivity of ~60 %. In the same patient the posttest probability drops from 26 to 21 % (where the LR+=60/1−0.78). This demonstrates that unless the test has very good diagnostic performance, it has a minimal effect when the pretest probability is low. If one were to start screening the general population, for example, all patients who pass through the ER with Strep tests alone, the pretest probability drops to 0.02 % and the posttest probability of a positive test drops to 0.07 %, which makes the test worthless to detect disease! All this is to make two points.

First is that knowledge of the patient population and the diagnostic performance of a test can help direct test orders; if a slow or relatively uninformative test may be avoided, there is potential to decreased length of patient stay and avoid the cost and risk of doing the test. Secondly, the laboratory can be a valuable source of the information regarding the diagnostic accuracy of tests to help with test selection and clinical decisions. Finally, inclusion of clinical criteria is essential, the laboratory provides but one essential part of the larger puzzle.

References

American Association for Clinical Chemistry Pediatric Reference Range Initiative. http://www.aacc.org/resourcecenters/resource_topics/pediatric_reference_range/

Baer, D. M., Ernst, D. J., Willeford, S. I., & Gambino, R. (2006). Investigating elevated potassium values. *MLO: Medical Laboratory Observer, 38*(11), 24. 26, 30–31.

Callum, G. (2001). *Fraser biological variation: From principles to practice*. Washington, DC: AACC Press.

Canadian Laboratory Initiative on Paediatric Reference Intervals (CALIPER). http://www.caliperdatabase.com/caliperdatabase/controller

Cappellini, M. D., Brancaleoni, V., Graziadei, G., Tavazzi, D., & Di Pierro, E. (2010). Porphyrias at a glance: diagnosis and treatment. *Internal and Emergency Medicine, 5*(Suppl 1), S73–S80. doi:10.1007/s11739-010-0449-7.

CLSI Document H3-A4. (1998). *Procedures for the collection of diagnostic blood specimens by venipuncture*. Approved standard, 4th ed. Wayne, PA.

Don, B. R., Sebastian, A., Cheitlin, M., Christiansen, M., & Schambelan, M. (1990). Pseudohyperkalemia caused by fist clenching during phlebotomy. *The New England Journal of Medicine, 322*(18), 1290–1292.

Ebell, M. H. (2003). Strep throat. *American Family Physician, 68*(5), 937–938.

Fox, J. W., Cohen, D. M., Marcon, M. J., Cotton, W. H., & Bonsu, B. K. (2006). Performance of rapid streptococcal antigen testing varies by personnel. *Journal of Clinical Microbiology, 44*(11), 3918–3922. Epub 2006 Sep 13.

Gaylord, M. S., Pittman, P. A., Bartness, J., Tuinman, A. A., & Lorch, V. (1991). Release of benzalkonium chloride from a heparin-bonded umbilical catheter with resultant factitious hypernatremia and hyperkalemia. *Pediatrics, 87*(5), 631–635.

Ji, J. Z., & Meng, Q. H. (2011). Evaluation of the interference of hemoglobin, bilirubin, and lipids on Roche Cobas 6000 assays. *Clinica Chimica Acta, 412*(17–18), 1550–1553.

Koch, T. R., & Cook, J. D. (1990). Benzalkonium interference with test methods for potassium and sodium. *Clinical Chemistry, 36*(5), 807–808.

Ma, Z., Monk, T. G., Goodnough, L. T., McClellan, A., Gawryl, M., Clark, T., et al. (1997). Effect of hemoglobin- and Perflubron-based oxygen carriers on common clinical laboratory tests. *Clinical Chemistry, 43*(9), 1732–1737.

National Institute of Health Children's Study. http://www.nationalchildrensstudy.gov/Pages/default.aspx

Sassa, S. (2006). Modern diagnosis and management of the porphyrias. *British Journal of Haematology, 135*(3), 281–292.

Van Steirteghem, A. C., & Young, D. S. (1977). Povidone-iodine ("Betadine") disinfectant as a source of error. *Clinical Chemistry, 23*(8), 1512.

Wolthuis, A., Peek, D., Scholten, R., Moreira, P., Gawryl, M., Clark, T., et al. (1999). Effect of the hemoglobin-based oxygen carrier HBOC-201 on laboratory instrumentation: Cobas integra, chiron blood gas analyzer 840, Sysmex SE-9000 and BCT. *Clinical Chemistry and Laboratory Medicine, 37*(1), 71–76.

FAQs

What is a lab error?, 3, 4, 17, 29, 31, 46, 59, 70

- This is an all-encompassing term for anything that goes wrong with a laboratory test from how it was ordered, how it was collected, how the sample was transported, how it was accessioned/received, how it was analyzed, and how it was reported. Laboratory errors often occur before the laboratory is involved in the process.

Why I can't use test _____ from that lab to compare with this result?, 70–71

- Different laboratories use different instrumentation, many laboratory methods are not harmonized; i.e., results differ and so do reference intervals used to interpret if they are abnormal.

Why did my reference interval change?, 89–91, 94

- Reference intervals often change with instrumentation. When a lab gets new hardware it is expected that some reference intervals will change to reflect the differences in test methods.

Why do you have to send the sample to a different lab?, 29, 79–81

- No single lab provides all possible tests. The less common the test, the more likely it is sent to another laboratory. This is largely a matter of resources.

What is a reference interval?, 2, 69, 72, 73, 77, 88–91, 94

- A reference interval represents the expected 95 % range of values obtained in a healthy population.

What are the limitations of reference intervals?, 2, 69, 72, 73, 77, 88–91, 94

- Defining a healthy population can be a challenge, consider pregnancy, age, common medications, and stature. Resources are also a challenge, where finding hundreds of healthy volunteers may be difficult, particularly at small laboratories.

R. Molinaro et al. (eds.), *Clinical Core Laboratory Testing*,
DOI 10.1007/978-1-4899-7794-6

How often does instrumentation change and why?, 61, 66, 69

- Typically laboratories change testing instrumentation every 5–8 years. Technology changes because technological improvements occur rapidly and there is substantial wear and tear incurred through running tests.

Why can't I have this test at the point of care?, 3, 20–22, 27, 28, 31, 46, 95

- POC testing in general has a limited menu and capabilities. It is difficult to make many assays perform robustly for POC; even many existing POC methods (e.g., glucose) do not perform as well as large automated methods. POC methods are often limited in scope where the results may not be valid in patients with extreme values are critically ill. Moreover, POC testing requires people using the test to continuously be certified as competent. POC testing is also more expensive than automated methods.

What test is done; where?, 1–23, 94

- Testing unicellular or multicellular and parasitic organisms are all part of microbiology. Common electrolyte, metabolite, proteins, enzymes, and drugs are part of clinical chemistry. Immunology testing includes auto-antibodies, such as ANA, ANCA, anti-paraneoplastic antibodies. Hematology covers blood cell testing (RBCs, WBCs, platelets) as well as coagulation. Overlap may occur with viral serology and hemoglobinopathies where a specialty that has the instrument to run a test may have it in their scope. Best option to answer this question is to find your local laboratory manual or call the laboratory and ask.

What color tube do I need for ____?, 4, 19–20, 35–50, 58, 77–95

- Different tube colors reflect different additives, which are essential for analysis. For example, red top tubes (containing clot activators) are designed to form clots, whereas green top tubes (containing heparin) are designed to prevent clotting. Each test has a preferred specimen type; for example, it is not possible to do coagulation testing on a clotted specimen. Your laboratory manual or laboratory staff (phlebotomy) can provide the information on the correct tube type. The wrong tube could yield no result or the wrong result; for example, a purple top tube (containing potassium-EDTA) will yield a very high potassium with a very low calcium).

Can I get a ___ on this body fluid (pancreatic cyst, drain fluid, etc.)?, 4, 19–20, 35–50, 58, 78, 87, 91, 93

- It is not possible to valid lab tests for every fluid. In general, body fluids are nonstandard, such that the composition of them may affect the results (high protein, variable pH, viscosity). Common fluids may be validated for a given test. There will not be available reference intervals to interpret these results as it is not possible to find healthy individuals willing/able to provide samples.

What's in this specimen?, 4, 19–20, 35–50, 58, 78, 87, 91, 93

- Labs don't have comprehensive methods to determine what is or is not in a specimen, particularly if the specimen is not from a human (e.g., what did my patient ingest). Labs may be able to test a specimen for specific analytes, which should be selected by the ordering physician (e.g., glucose, sodium, potassium, chloride, creatinine). If it is a question between one fluid and another, a consultation with a clinical chemist is likely to be beneficial.

Will you convert these units for me?, 9, 14, 38, 46, 77–78, 94

- Yes! Laboratorians are uniquely equipped to convert between convention and nonconventional units.

Why is the test taking so long?, 37, 58, 77–95

- Depending on which test it is, results may take up to a month to be reported. If it's something that is usually fast, it is worth calling the laboratory to determine if the sample was received or if there is some identifiable cause for delay. Labs should be able to provide a reasonable estimate of how long any given test takes.

Why is the lab calling?, 37, 58, 77–95

- Laboratories have policies and procedures in place designed to provide timely information to those caring for patients. This may include critical values, abnormal results, or other flags that warrant a conversation. The lab is blind to the status of the patient such that these calls may seem like a nuisance. For example, the lab may call about extreme results when the patient is critically ill and the healthcare team knows it.

Why was the sample rejected?, 4, 19–20, 35–50, 58, 78, 87, 91, 93

- To yield accurate results, samples must meet minimal acceptance criteria. These include the right tube type, collected at the right time, transported within a given timeframe at a defined temperature. Labeling is essential where there must be identifiers for both patient and ordering physician.

What causes hemolysis/icterus/lipemia?, 4, 19–20, 35–50, 58, 77–95

- Hemolysis results from breaking red cells and may occur after the sample is collected or within a patient. Icterus results from very high bilirubin and occurs endogenously. Lipemia occurs when lipids are very high, but can also be found in contaminated collections from drug infusions, such as propofol. When these conditions occur endogenously, there are limited options for recollection or the lab to obtain a result. When hemolysis of IV contamination occurs, a fresh sample collection is the only option. Those who draw samples should know that the lab cannot cause any of these conditions. Dropping a sample doesn't cause hemolysis, but collecting it through a small needle can.

What's the best test to diagnose ____?, 1–23, 25–34, 94

- Where questions about which test to order arise it is useful to consult a clinical chemist for chemistry tests, a microbiologist for microbiology tests, and a hematologist for coagulation, blood products, and transfusion medicine. These individuals are all available on call 24/7 and can be contacted through hospital paging systems or laboratory information lines.

Who should I call to ask about test ____?, 1–23

- As above, it is useful to consult a clinical chemist for chemistry tests, a microbiologist for microbiology tests, and a hematologist for coagulation, blood products, and transfusion medicine. These individuals are all available on call 24/7 and can be contacted through hospital paging systems or laboratory information lines.

How does drug/food/herbal product affect ____?, 43, 61, 62, 66, 69, 73, 74, 75, 85–87

- Clinical chemists have numerous resources at hand and can provide information about interference from different compounds, nutritional supplements, and herbal remedies. However, it is often difficult to find reliable information for a given compound for a given assay, particularly when there are numerous metabolites.

What volume do I need to collect for _____?, 4, 19–20, 35–50, 58, 78, 87, 91, 93

- Laboratory manual typically provide detailed information about how to collect a sample for a given test. If that information is unavailable, the lab is always prepared to provide this information verbally.

What is the formula for corrected calcium, eGFR, etc.?, 2, 3, 9–14, 17, 19–22, 37, 53, 57–59, 61, 65, 66, 69, 71, 76, 92, 93

- The laboratory manual may or may not provide detailed information about calculated values. A call to the clinical chemist is certain to yield answers to such questions.

How do I follow up with bizarre test result from outside lab (hair antimony, urine uranium, etc.)?, 3, 25–34, 53, 79–81

- Laboratory medical/scientific staff are useful resources for interpreting strange results. They are also able to provide information about whether a source lab is reliable and accredited.

Why is drug screen negative/positive when patient is prescribed ____?, 2, 3, 9–14, 17, 19–22, 37, 43, 53, 57–59, 61, 62, 65, 66, 69, 71, 73–76, 85–87, 92, 93

- Urine drug screens are often less specific for drugs of interest that desired. Amphetamines and opiate assays are notorious for cross-reactivity, while at the same time potentially missing related compounds, such as oxycodone. Consultation with a clinical chemist is useful as there are confirmation tests available and chemist will be aware of cross-reactivity rates with different substances. It is also essential to recognize that a laboratory can't tell when a drug was taken or how much was taken, it is simply a snapshot of in-time.

I read about a test in a paper, how do I order it?, 1–23, 25–34, 80, 94

- There are many tests that have been researched that never make their way into the clinical laboratory. This is due to a variety of reasons from regulatory hurdles, cost, reliability, to reproducibility and the availability of reference intervals.

Why are genetic/molecular tests so expensive?, 3, 25–34, 80

- Molecular tests use expensive hardware and reagents and are usually done in low volumes. In addition, they are done at reference laboratories, all of which drive up costs.

Index

R. Molinaro et al. (eds.), *Clinical Core Laboratory Testing*,
DOI 10.1007/978-1-4899-7794-6

Printed in the United States
By Bookmasters